Sustainable Production, Life Cycle Engineering and Management

Series editors

Christoph Herrmann, Braunschweig, Germany
Sami Kara, Sydney, Australia

Modern production enables a high standard of living worldwide through products and services. Global responsibility requires a comprehensive integration of sustainable development fostered by new paradigms, innovative technologies, methods and tools as well as business models. Minimizing material and energy usage, adapting material and energy flows to better fit natural process capacities, and changing consumption behaviour are important aspects of future production. A life cycle perspective and an integrated economic, ecological and social evaluation are essential requirements in management and engineering. This series will focus on the issues and latest developments towards sustainability in production based on life cycle thinking.

More information about this series at http://www.springer.com/series/10615

Rainer Stark · Günther Seliger
Jérémy Bonvoisin
Editors

Sustainable Manufacturing

Challenges, Solutions and Implementation
Perspectives

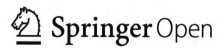 Springer Open

Editors
Rainer Stark
Chair of Industrial Information Technology,
  Institute for Machine-tools and Factory
  Management
Technische Universität Berlin
Berlin
Germany

Jérémy Bonvoisin
Chair of Industrial Information Technology,
  Institute for Machine-tools and Factory
  Management
Technische Universität Berlin
Berlin
Germany

Günther Seliger
Chair of Assembly Technology and Factory
  Management, Institute for Machine-tools
  and Factory Management
Technische Universität Berlin
Berlin
Germany

Editorial assistants:
Paul Einhäupl (project coordination)
Helena Roth (graphic post-processing)

ISSN 2194-0541   ISSN 2194-055X   (electronic)
Sustainable Production, Life Cycle Engineering and Management
ISBN 978-3-319-83960-8   ISBN 978-3-319-48514-0   (eBook)
DOI 10.1007/978-3-319-48514-0

Printed on acid-free paper

This Springer imprint is published by Springer Nature
The registered company is Springer International Publishing AG
The registered company address is: Gewerbestrasse 11, 6330 Cham, Switzerland

Foreword

Manufacturing is the main driver for welfare and prosperity of people. However, manufacturing also strongly contributes directly and indirectly to the depletion of natural resources, environmental burdens—affecting the health of animals, humans, and eco-systems—as well as to social conflicts. These negative effects expand along with the worldwide demand for industrial goods, which will further increase since the global population is still growing and less developed countries strive for the standard of living which richer countries already have achieved. And despite being aware of the prevalent limitations of natural resources and emission capacities of our planet, the demand for resources and the related pollution to the environment has continued to rise drastically. Thus, finding solutions towards a more sustainable development of global manufacturing—which simultaneously considers the triple bottom line with the three dimensions of sustainability—is of outmost importance and more urgent than ever.

Researchers of the CRC for Sustainable Manufacturing have taken up the resulting challenges and derived ambitious goals. These goals address the main tasks supporting the shift towards a more sustainable development in manufacturing, which are the identification of challenges and levers for change which matters most, the development of specific solutions to cope with these challenges, and the implementation of decision support methods for supporting deciders in the industry and policy to improve manufacturing activities based on the derived solutions. The identification of challenges requires a system thinking and a life cycle orientation in order to avoid problem shifting. Specific improvement measures may lead to local improvement but create negative effects on other manufacturing sectors, life cycle phases, or environmental impact categories. That means it is necessary to consider and evaluate different manufacturing scales from specific technologies and product concepts to value creation networks and up to global manufacturing activities regarding economic, environmental, and social criteria. Furthermore, innovative solutions must be found to improve technologies, products, and strategies for manufacturing activities which reduces resource demands, create products in the desired quality, and protect the health of workers, customers, and all involved people. These solutions must be transformed into methods and tools—such as mathematical optimization approaches for specific planning problems or life cycle assessment models for products and processes—which supports engineers, planners, and designers in creating value-adding and sustainable products and services.

This publication provides research results which address the aforementioned challenges, solutions and implementation perspectives with regard to manufacturing and sustainable development. It contributes to this urgent topic by describing prevailing trends and findings in industry, economy, and society, by presenting concepts for manufacturing technologies, planning methods, and product designs, as well as by suggesting strategies for knowledge dissemination and employment of solutions within organizations. Overall, the present book gives insight into important fields of actions which are required to make the world worth living in now and for future generations.

Technische Universität Braunschweig   Prof. Dr.-Ing. Christoph Herrmann
University of New South Wales   Prof. Dr. Sami Kara

Preface

If the lifestyles of up-and-coming and also developed societies are shaped in the future by the existing, currently predominating technologies, then the resource consumption at play will exceed every accountable environmental, economic and social boundary known to man. The dynamics of global competition and cooperation can be utilized for lending wings to processes of innovation and mediation towards the ultimate goal and necessity of *sustainability* on our globe. A special focus lies in condensing engineering to *sustainable manufacturing*, thus specifically addressing artefact generation shaping human living.

Abstract, intangible concepts and goals, such as *sustainability,* overburden human beings, engineers and researchers in different ways. To date, it constitutes an overwhelming task to consistently apply a full, balanced view and critical cross-assessment of the full range of relevant dimensions, such as the environment (incl. climate, resources and all other natural systems), the economy and society—the classical three pillars of the modern understanding of sustainability. Furthermore, conducting such assessments on different levels of abstractions creates a personal feeling of powerlessness. Researchers, however, take on the challenge of investigating both laws of interdependence and the underlying core mechanisms in order to provide new systemic views of the challenging concepts and goals of *sustainability*. Engineers, in addition, attempt to derive methods, processes and technologies to help society and companies in finding holistic, specific and proactive solutions for *sustainability*. In that mission, *sustainability* gets broken down into controllable elements within an overall system network: products with their functions and behaviours, material selections, production systems, factories, enterprises, logistic elements, value creation networks, patterns of use behaviours, labour and payroll systems, welfare, health and so forth.

The editors of this book and all contributing authors are of the belief that it is now high time to provide tangible solution sets to address various levels of "system driven realization and delivery oriented" sustainability. Yet what does that actually entail? Unlike the rather general (but necessary) discussions surrounding a "complete enough set" of sustainability development goals, such as the 17 goals agreed by the United Nations in September 2015, a critical urgency is attached to work on "how such goals need to be realized, and which solutions have to be made available for them." Manufacturing and its potential to deliver wealth and intelligent solutions to societies and human beings has been selected as a first prime route for investigating which changes are necessary for reaching a true state of *sustainability* in the future.

Sustainable manufacturing in this sense represents a manufacturing engineering's approach to coping with these challenges. Manufacturing technology is developed in the direction of economic competitiveness, of environmental compatibility with natural global frame conditions of resource availability and of social welfare with different societal frame conditions to suit the different human communities around the globe. This multidimensional goal system can be balanced by developing adequate economic, environmental and social criteria, with analysis of their interdependencies and application of that analysis for guiding technological innovation in respective economic, environmental and societal frameworks.

Before stepping up and striving for a position from which to set up a "circular economy"—an economy where the value of products, materials and resources is maintained in the economy for as long as possible, and the generation of waste is minimised—it is necessary to determine which elements need to be integrated into such a "system circle." Hence, the underlying new mind-set of this book assumes that the overall values of *sustainability* and those specific to *sustainable manufacturing* can no longer just be driven by assessment factors and goals of the three dimensions environment, economy and society. Instead, it has become necessary to fundamentally change core and specific mechanisms and elements of the following "interacting system of systems":

- The earth system with its natural resources and all associated ecological, biological and climatic sub-systems
- The societal system(s) and related behavioural patterns which are highly influenced by cultural, religious and ethnic values
- The economic system(s) originally driven by a profit theory based on the traditional production factors, such as land, work and capital, funds which was just recently however hugely impacted by new business model innovations, e.g. caused by the digital transformation

Research efforts in this book have investigated both specific technical approaches for ushering in changes to specific mechanisms ("the technical depth"), and on rather generic terms, overarching theories and methodologies on how value creation and its technical solutions can be variously influenced by specific earth and economic boundary conditions, i.e. the breadth of the overall approach.

The rather limited predictive capability of the timely progression of evolutionary or revolutionary changes to the solution set of *sustainable manufacturing* remains the first generic challenge of research in this field. This is the reason why system dynamic models are deemed to be appropriate candidates for overcoming such a research dilemma.

The second generic challenge in *sustainable manufacturing* deals with the contradiction between

- the desire to analyse the different aspects of the lifecycle behaviour of a product as narrowly as possible, and
- the obligation to provide a rather lean set of data, information and digital models for the design and determination of a product solution in the "Begin of Life" (BOL) phase of the lifecycle.

The right mix of both is decisive in the pursuit of enhancing the probability of influence on "smart and comprehensive decisions" as part of the engineering process of sufficiently sustainable products.

The ambition behind and need for driving changes from the different development levels of society and economy within both the highly developed and industrialized countries as well as from the perspectives and demands of the less developed and emerging countries represent the third overall challenge in *sustainable manufacturing*. Positive impact of the manufacturing sector on sustainability will thus only be possible if all participants actively involved think *locally and globally*.

This book is unique in its comprehensiveness in tackling research and engineering approaches in *sustainable manufacturing* and its global value creation mechanisms. It is the desire and intention of the editors that this book may serve to truly help researchers, industrial experts, politicians and interested members of society in the process of fully comprehending and further developing new solutions for driving and realizing *sustainability* with the help of manufacturing solutions. It is therefore an obligation for editors to stay in close contact with the growing community of sustainability oriented researchers, planners, engineering, managers, politicians and responsible individuals in all societies across the world.

Rainer Stark
Technische Universität Berlin

Contents

Part I
Challenges

Field of Research in Sustainable Manufacturing

Jérémy Bonvoisin, Rainer Stark and Günther Seliger

Abstract Sustainability has raised significant attention in manufacturing research over the last decades and has become a significant driver of the development of innovative technologies and management concepts. The current chapter aims to provide a structured overview of the wide field of research in sustainable manufacturing with a particular focus on manufacturing technology and management. It intends to describe the role of manufacturing in sustainability, outline the complementary approaches necessary for a transition to sustainable manufacturing and specify the need for engaging in interdisciplinary research. Based on a literature review, it provides a structuring framework defining four complementary areas of research focussing on analysis, synthesis and transition solutions. The challenges of the four areas of research manufacturing technologies ("how things are produced"), product development ("what is being produced"), value creation networks ("in which organisational context") and global manufacturing impacts ("how to make a systemic change") are highlighted and illustrated with examples from current research initiatives.

1 The Role of Manufacturing in Sustainability

Humanity is increasingly confronted with the challenge of dealing with a finite earth—a world with a limited "carrying capacity" (Arrow et al. 1995) and with "planetary boundaries" (Rockström et al. 2009), with some expecting "limits to growth" (Meadows et al. 1972). Owing to the unprecedented growth in population and economic output experienced since the 19th century (respectively six and sixty-fold, Maddison 2006), the stress imposed by humanity on natural equilibria

J. Bonvoisin (✉) · R. Stark
Chair of Industrial Information Technology, Institute for Machine-tools
and Factory Management, Technische Universität Berlin, Berlin, Germany
e-mail: bonvoisin@tu-berlin.de

G. Seliger
Chair of Assembly Technology and Factory Management, Institute for
Machine-tools and Factory Management, Technische Universität Berlin,
Berlin, Germany

© The Author(s) 2017
R. Stark et al. (eds.), *Sustainable Manufacturing*, Sustainable Production,
Life Cycle Engineering and Management, DOI 10.1007/978-3-319-48514-0_1

3

has reached alarming levels at the same time that it fortifies increasing inequality between early industrialised and emerging countries. The limited capacity of the atmosphere to take stock of the emissions produced by our carbon-based economies, poses a threat not only to natural equilibria, but also to our own daily conditions of living (Edenhofer et al. 2015). The flows of some elements due to human activities, such as phosphor and nitrogen, now exceed natural flows, thus threatening the balance of the metabolism of natural ecosystems (Vitousek et al. 1997). Hence, the risk of "overshooting", i.e. drawing on the world's resources faster than they can be restored, while releasing wastes and pollutants faster than the earth can absorb them, is very real and the ongoing, unresolved challenge of our time (Meadows et al. 2004).

Although the concept of "sustainable development" (as defined for example by Brundtland et al. 1987) has received significant attention and motivated numerous initiatives in favour of, e.g. recycling, energy efficiency, the need for action is now nevertheless greater than ever before. This is particularly underscored by the observation that, despite international efforts to combat climate change, the global energy system is carbonizing due to a global renaissance of coal (Steckel et al. 2015). Further and more innovative decarbonisation solutions are therefore urgently needed.

As a major stakeholder in several areas of human living, industry has a great role to play in sustainability. It first contributes significantly to the overall environmental impact of human activity. It represents 26 % of the final energy consumption in the EU 27 (Lapillonne et al. 2013, data from 2013), emits 28.5 % of the greenhouse gases produced in the EU 27 (European Commission 2013) and uses energy which is still generated from fossil energy sources by up to 56 % (Lapillonne et al. 2013, data from 2013). In 2006, the European Commission estimated an overall European energy saving potential of 20 %. In the case of industries, the potential savings are estimated to be 25 %, representing annual losses of about 100 billion euros (European Commission 2006). At the same time, while the precision of production processes reaches ever smaller scales, the energy consumption of corresponding production systems is increasing exponentially (Gutowski et al. 2011). Meanwhile, further increases in energy consumption are anticipated.

Beyond its direct environmental impacts, the discrete product manufacturing sector also influences the resource consumption of its products over their entire lifecycle, and therein plays a critical and complex role in sustainability (Duflou et al. 2012). This role is particularly relevant considering that households in early industrialised countries face a literal "rise of the machines" and are equipped with more products and appliances than only a few decades ago (Energy Saving Trust 2006). The average household in early industrialised countries may own thousands of material items, so managing the volume of the possessions becomes a stress factor (Arnold et al. 2012).

With respect to the social aspects, the industrial sector employs 17 % of the European workforce (Eurofound 2012) and represents more than 23 % of worldwide total employment (International Labour Organization 2014). On the other hand, while working conditions in the manufacturing sector have improved steadily over the last decades (World Health Organization 2013), poor working conditions persist in resulting in as many as 300,000 work-related deaths and economic losses of 4 % of the gross domestic product of the European region every single year (WHO 2016). Globally, industries are responsible for 7.2 % of child labour, or 12 million people (Diallo et al. 2013).

That said, manufacturing stands strong as a crucial sector for the development of economies. Manufacturing generates 14 % of the gross domestic product (GDP) of OECD countries and of Europe according to the OECD (2016),[1] and 31 % of the world GDP according to the US central intelligence agency (2016).[2] Beyond this quantitative contribution to the GDP, whose reflection of actual wealth is debatable (see e.g. Costanza et al. 2014), it has been shown that stable specific and sequential sectoral patterns can be observed in economic development processes across the spectrum of countries, with specific manufacturing sectors furthermore playing an important role in initializing economic development processes in poor countries (Radebach et al. 2014). On the whole, thus, basic manufacturing activities seem to be a necessary enabler for the development of modern economies.

To summarize, manufacturing as a subset of the industrial sector (see glossary for disambiguation of the terms) has a threefold impact on sustainability:

- it plays a major role in the creation of wealth;
- it directly contributes to the material metabolism of human societies as it requires material input and produces outputs;
- it indirectly contributes to the material metabolism of human societies as it produces outputs having their own metabolism even after having left manufacturing systems.

2 Existing Approaches of Sustainable Manufacturing

As a counterpoint to this tripartite observation, sustainable manufacturing is defined in the present publication as (see also the glossary for more information on this definition):

> creation of discrete manufactured products that in fulfilling their functionality over their entire life cycle cause a manageable amount of impacts on the environment (nature and society) while delivering economic and societal value.

The international research community has been particularly active in the last decades in the development of conceptual or concrete solutions toward sustainable manufacturing (see for example Arena et al. 2009). The objective of the current contribution is to deliver a framework for providing a structured overview of the existing field of research in sustainable manufacturing, with a particular focus on industrial engineering. It intends to outline the complementary approaches required

---

[1]Accessed 09.03.2016. Figures for EU-28/2015 and for OECD/2014.
[2]Accessed 22.08.2016, last updated 04.02.2016.

for a transition to sustainable manufacturing and their necessary interdisciplinary modus operandi. While Sect. 2.1 provides an overview of previous attempts in this direction, Sect. 2.2 introduces an original framework of sustainable manufacturing, according to which the present book publication is structured. Section 3 is specifically dedicated to the discussion of the challenges of multi-, inter- and transdisciplinary approaches faced by researchers in sustainable manufacturing.

2.1 Review of Published Frameworks

Since the emergence of the first initiatives explicitly termed as green engineering or sustainable manufacturing, several reviews of the field have been undertaken and frameworks have been proposed that identify the complementary areas of research that need to be addressed. Jayal et al. (2010), for example, deliver an overview of strategies for sustainable manufacturing with a particular focus on the modelling and assessment techniques for the development of sustainable products, processes and supply chains. Duflou et al. (2012) provide an extensive review of strategies for energy and resource efficiency in discrete part manufacturing, considering five complementary levers: unit process, manufacturing line, facility, manufacturing system and global supply chain. Based on the evaluation of the potential of these techniques, they estimate potential energy savings of 50 % in the overall consumption in the manufacturing sector. Garetti and Taisch (2012) furthermore published an overview of trends affecting the manufacturing sector, highlighting the challenges raised by sustainability in this sector and the corresponding strategies. They identify four complementary research clusters with a broader focus: enabling technologies, resources and energy management, asset and product lifecycle management, business model and processes. Finally, Haapala et al. (2013) made recommendations for further research on sustainable manufacturing, based on the review of existing initiatives and considering two foci: manufacturing processes and equipment along with manufacturing systems.

It is worth noting that all these reviews identify both sustainability assessment methods and technical strategies (*analysis* and *synthesis*) as necessary and complementary approaches to sustainable manufacturing. Analytical approaches are required in order to put words and figures to the problems which may ultimately be solved by synthesis. One example of this is found in the inventory of approaches for energy efficient manufacturing at the unit process level given by Duflou et al. (2012), where data acquisition, computational models and energy assessment methods stand alongside technical solutions such as "technological change" or "waste recovery within the machine tool." Two of the four publications go further, and state that analysis and synthesis approaches can only be effective if enabled by adapted *education* tactics. On one side of the equation, a systematic implementation of analysis and synthesis approaches in industry requires that engineers fully appreciate the sustainable manufacturing concepts and are trained in multi-objective decision-making. On the other side, the general public can only foster sustainable production if they fully appreciate the impact of their consumption patterns.

While such reviews identify different yet overlapping scopes, the sustainable manufacturing solutions they identify can be classified into four different areas, which we will call for our purposes *layers*:

- Manufacturing technologies: approaches focused on "how things are manufactured", i.e. whose object of research lies in processes and equipment, including machine-tools or facilities. Examples of such approaches are among other things: development of new or improved manufacturing processes, predictive maintenance of production equipment, determination of process resource consumption, process chain simulation, or energy-efficient facility building.
- Product lifecycles: approaches focussed on "what is to be produced", i.e. whose object of research is the product definition (where product can be understood as a good or a service). Examples of such approaches are among others: asset and product lifecycle management, intelligent product, simplified product sustainability assessment.
- Value creation networks: approaches focused on the organisational context of manufacturing activities, i.e. whose objects of research are organisations such as companies or manufacturing networks. Examples of such approaches are among others: resource efficient supply chain planning, industrial ecology.
- Global manufacturing impact: approaches focused on the transition mechanisms towards sustainable manufacturing, i.e. whose objects exceed the conventional scope of engineering. Examples of such approaches are among others: development of sustainability assessment methods, education and competence development, development of standards.

Table 1 summarizes how the four cited reviews of the field of sustainable manufacturing correspond to the four identified layers.

Table 1 Four layers of sustainable manufacturing identified in previous frameworks

Layer	Object addressed	Haapala et al. (2013)	Garetti and Taisch (2012)	Duflou et al. (2012)	Jayal et al. (2010)
Global manufacturing impact	World (society, environment, economy)	•	•		
Value creation networks	Organisations (companies and manufacturing networks)	•	•	•	•
Product lifecycles	Product definition (good and service)		•		•
Manufacturing technologies	Process and equipment (machine-tool, facility)	•	•	•	•

As a last observation, it should be noted that although these reviews define sustainable manufacturing as resulting from the consideration of the three dimensions, the specific solutions which they present remain confined to the environmental dimension (or even consider resource efficiency exclusively) and in so doing, elude the social dimension altogether. This is in accordance with the observation provided by Arena et al. in 2009 already, in their extensive state-of-the-art of industrial sustainability study: while the social dimension of sustainability is generally viewed to be worth considering, only few specific solutions have been provided to date which address these social issues. In their summary of published research on the role of manufacturing in social sustainability, Sutherland et al. (2016) state that manufacturing enterprise still lacks standardised approaches for internalising social sustainability and for outlining directions of future work in order to mitigate this situation, such as the further development of Social Life Cycle Assessment (S-LCA).

Based on these contributions and the observations made, the next section introduces a framework structuring the field of the necessary research for enabling the transition to sustainable manufacturing.

2.2 Proposed Framework

Manufacturing activities can be characterised as the interplay of five value creation factors, i.e. human, process, equipment, organisation and product, taking place in value creation modules (Seliger et al. 2011). Value creation modules are, in turn, vertically and horizontally integrated into geographically distributed value creation networks. Value creation modules generate effects on the three dimensions of sustainability that can be measured by sustainability assessment methods.

Following the value creation network model depicted in Fig. 1 and based on the findings of the previous section, sustainable manufacturing can be defined as the necessary interplay of three kinds of approaches:

- analysis approaches, i.e. methods allowing the evaluation of value creation based on the three dimensions of sustainability;
- synthesis approaches, i.e. implementation of these methods in the development of technical systems at all levels of value creation (value creation factors, modules and networks);
- approaches for systemic changes, i.e. to transform business to become standard vehicles towards sustainable processes; in other words: enabling the systematic integration of sustainability in day-to-day decision-making.

These approaches are embedded in the four concentric and sequentially including areas introduced in the previous section: manufacturing technologies, product lifecycle, value creation networks, global manufacturing impact. The interplay of analysis, synthesis and transition approaches and these four layers are

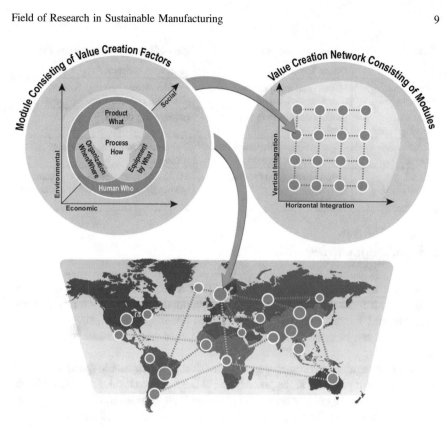

Fig. 1 Value creation network (VCN) model

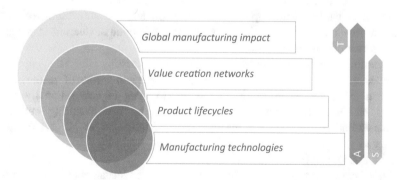

Fig. 2 Interplay of analysis, synthesis and transition approaches and the four areas of sustainable manufacturing (*T* transition; *A* analysis; *S* synthesis)

depicted in Fig. 2 while Table 2 presents their respective scientific disciplines and objects of research. Layers are depicted with more detail in the subsequent sections of this chapter.

Table 2 Objects and scientific disciplines of the four layers of sustainable manufacturing

Layer	Object addressed	Discipline concerned
Manufacturing technology	Process and equipment (machine-tool, facility)	Production engineering, factory planning, operation management
Product development	Product definition (good and service)	Engineering design
Value creation networks	Organisations (companies and manufacturing networks)	Business economics, knowledge management
Global manufacturing impact	World (society, environment, economy)	Micro and macro-economics, natural sciences, humanities, politics, education

2.3 Manufacturing Technologies

This layer specifically addresses the two factors of value creation *process* and *equipment*. It focuses on the development of production technologies, machine-tool concepts and factory management techniques ensuring that whatever has to be produced, it can be done with economy of resources which likewise uphold social standards.

This first requires determining specific indicators which enable the identification of improvement potential at the process and at the machine level. Examples of these are found in the "specific energy consumption," an empiric model developed by Kara and Li (2011) for material removal processes and based on measures on machine tools, or the "electrical deposition efficiency," an analytic model developed by Sproesser et al. (2016) for welding processes. At facility level, cyber-physical systems (Low et al. 2005) and metering techniques (Kara et al. 2011) can be employed in tandem with appropriate facility models and simulation techniques (e.g. Herrmann and Thiede 2009) in order to enable optimal steering of processes within a manufacturing system.

Regarding the development of new technologies, existing efforts encompass, for example, the improvement of welding technologies in terms of resource consumption (Sproesser et al. 2015) or the development of new internally cooled cutting processes (Uhlmann et al. 2012). At the manufacturing cell level, lifetime-extending add-ons for machine-tools (Kianinejad et al. 2016) and of automated workplaces preventing musculoskeletal strain by workers (Krüger and Nguyen 2015), can be cited as examples.

While such solutions form a necessary basis for sustainable manufacturing, macroeconomic calculations underscore that applying best available sectorial technologies in all regional industry sectors across the world would reduce CO_2 emissions to one-third (Ward et al. 2015). This shows that solutions are required beyond the manufacturing technology level in order to reach e.g. the factor 4 or 10 pinned by some authors as a necessary objective of environmental reduction of human activities (e.g. Weizsacker 1998).

This layer is specifically addressed in the part "Solutions—Sustainability-driven Development of Manufacturing Technologies" of the present book.

2.4 Product Lifecycles

This layer specifically addresses the factor of value creation *product*. It focuses on enabling the operation of product development processes systematically leading to products which achieve balance of the three dimensions of sustainability, i.e. which generate low environmental impacts while delivering socially useful functions, all available at reasonable production and purchase prices. This requires the application of methods allowing product development teams to systematically integrate sustainability criteria into their decisions.

Over the past decades, a large variety of methods of this type have been developed. As early as 2002, Baumann et al. identified more than 150 methods for "green product development", i.e. focusing strictly on the environmental dimension of sustainability, while Pigosso (2012) more recently identified 106 of them. The wide range of methods generated by the scientific community led Ernzer and Birkhofer (2002) to state that the difficulty no longer lies in developing design methods, but lies rather in selecting the relevant methods and applying them efficiently. As a matter of fact, existing methodological support for sustainable product development is often criticized for being poorly integrated into the product development process, ultimately leading to additional exertion on the part of product development engineers, and at the same time to low industry diffusion (Rosen and Kishawy 2012; Knight and Jenkins 2009).

Addressing this very issue, Pigosso et al. (2013) developed a maturity model which allows a step-by-step, guided integration of sustainable product development methods in companies. At a more operational level, Buchert et al. (2014) developed an IT-tool aimed at supporting the selection of the appropriate method for a given design problem. From the flipside of the process, some other authors have striven to reduce the diversity of tools through the development of integrated frameworks (e.g. Dufrene et al. 2013). In all cases, a key factor for effective consideration of sustainability in daily product development activities is found in the integration of methods in information systems such as Product Lifecycle Management (Stark and Pförtner 2015).

Given the high number of constraints applying to product development which limit the solution space spectrum along with the attainable level of innovation, parts of the research community have striven to reclaim degrees of freedom in their pursuits, by fostering alternative production or consumption patterns. A well-researched topic in this area is found in the concept of product service systems through which: "it is in the economic and competitive interest of the producer/provider to foster continuous innovation in reducing the environmental impacts and improving social equity and cohesion" (Vezzoli et al. 2015). Another partially overlapping field of research is found in the participative design models allowing for a deeper integration of the voice of the final user in the design process, such as user-centred design or open source design (Aitamurto et al. 2015; Bonvoisin and Boujut 2015).

This layer is specifically addressed in the part "Solutions—Sustainable Product Development" of the present book.

2.5 Value Creation Networks

This layer addresses the value creation factor *organisation* as well as the combination of value creation modules into value creation networks. It addresses the ability of the value creation networks to support sustainable production and products. How sustainable a product proves to be, may, for instance, be determined not only by its design, but also by an array of choices made in the value creation network that are not accessible to the product development team. More specifically, a given product cannot be claimed to be sustainable universally or inevitably, but in relation to a given context and associated use (Manzini and Jégou 2003). The remanufacturability of a product, furthermore, only constitutes potential that is born out of the product design itself, and can only be realized by the interplay of activities including, among other things, reverse logistics, product dismantling and testing. How sustainable a transportation system based on electric cars proves to be for a given area, for example, may depend on the density of the population and the existence of an appropriate public transportation network. Following Haapala et al. (2013) in that pursuit, then, the question lies not only in which processes are performed, but also *where* these processes are performed. This question is notably important in a world of globalized supply chains where intensive processes tend to be outsourced to emerging countries (Andersson and Lindroth 2001; Bonvoisin 2012).

Taking this into consideration, approaches are required to help ensure the development of organisational infrastructure which facilitates sustainable products and productions. Two critical aspects identified by Jayal et al. (2010) are multi-objective and integrated value creation planning. One challenge lies in moving from the coordination of independently managed organisations with individual profit maximisation behaviour, to more integrated planning. The other challenge is to go beyond profit minimisation and integrate several dimensions into the decision-making process in pursuit of connecting value creation modules.

This layer is specifically addressed in the part "Solutions—Sustainable Value Creation Networks" of the present book.

2.6 Global Manufacturing Impact

This last layer addresses the penetration rate of sustainable solutions, i.e. how far sustainable decision-making methods are implemented in practice. In order to pave the way for necessary cultural change, research which takes on the triple role of yardstick (measuring sustainability), guidepost (setting targets) and multiplier (motivating towards a direction), is what is required.

The first role requires the development of methods for measuring the actual sustainability performance of products and manufacturing activities, examining improvement potentials and identifying trade-offs between the achievement of multiple targets. As a central methodology in sustainable engineering, Life Cycle Assessment (LCA) and even more relevant, Life Cycle Sustainability Assessment (LCSA) (Finkbeiner et al. 2010), figure as essential parts of the solution. These tools however represent heavy machinery that remain too time-consuming and difficult for engineers to appreciate, and therefore hardly applicable in day-to-day decision-making. In particular, a first task lies in equipping engineers with the knowledge and framework of reference necessary to select appropriate indicators among the huge amount of indicators available. A second predicament underlined by Jaya et al. (2010) lies in the development of rapid and convenient sustainability evaluation procedures which yield results as precise as LCA.

The second role requires the development of methods for setting appropriate sustainability targets. For example, most LCA indicators (e.g. global warming potential) have been primarily developed for determining the sustainability performance of a product or process in comparative terms (i.e. in comparison with another product or process delivering the same function). Hence, they can support manufacturing that always strives to "be more sustainable than before" but cannot ensure that manufacturing is sustainable in absolute terms (Bjørn and Hauschild 2013). Yet, despite however useful they may be for comparing processes or products, these indicators need to be complemented by a sustainability analysis in more absolute terms. This includes both the setting of clear sustainability reference values/targets (e.g. maximum allowed CO_2 emissions to meet the $2°$ goal) and the development of methods to analyse the sustainability of products and processes with regard to these targets (as proposed by Bjørn et al. 2016, for example).

The third role involves the overall effort attached to the information transfer to industry, policymakers and the general public, in order to stimulate the necessary cultural change. One essential lever in that pursuit advocated by Haapala et al. (2013), Mihelcic et al. (2003) and Garetti and Taisch (2012) is non other than pure and simple education. On the one hand, manufacturing-related curricula should provide engineers with a broader understanding of the concept of sustainability and of the influence of their activities on societal and environmental systems. They should be able to identify improvement potential in technical systems towards sustainability, evaluate optimal solutions, and take decisions accordingly. At the same time, they should be made to appreciate the socio-technical nature of sustainable manufacturing, along with the influence of the behaviour of consumers and users on the other side of the spectrum. On the other hand, the actual transition towards sustainability not only relies on engineers, but also on the "environmental" and "technological literacy" (Mihelcic et al. 2003) of the greater citizenry, which would allow people to make enlightened and balanced consumer decisions. Considering empirical observations showing that both concepts of sustainability and manufacturing may not generally be well understood (e.g. Roeder et al. 2016), a tremendous need is present for the integration of all such concerns in education agendas, from primary school to university.

This layer is specifically addressed in the part "Implementation Perspectives" of the present book.

3 Challenges of Interdisciplinarity in Sustainability Research

The above detailed layers are not only complementary on the topics which they address, but likewise interdependent. Stock and Burton (2011) note that sustainability "necessitate[s] solutions informed by multiple backgrounds that singular disciplines seem unable to provide, and possibly, are even incapable of providing" and therein they underline the necessity for collaboration between the disciplines. They differentiate between multi- and interdisciplinarity: while multidisciplinarity is characterized by the co-existence of different scientific disciplines with parallel objectives in a common research field, interdisciplinarity seeks to bridge disciplinary gaps in perspective by involving different disciplines in the achievement of a common goal. Together with Schäfer (2013), they even advocate for transdisciplinary research, i.e. the inclusion of non-researcher stakeholders such as representatives from enterprises, administration or NGOs, end-users or citizens in the process of producing solutions of complex socio-technical problems. One argument for this is that the very concept of sustainability cannot be stated universally, but instead has to be considered within each and every specific social context. This requirement is backed by the strong observation stressed by Mihelcic et al. already in 2003 that engineering disciplines lack connective oversight of societal problems, that the public has difficulty appreciating what exactly engineers do, and that engineers tend to overlook the social dimension attached to the socio-technical problems which they invariably address. A further tendency to isolation of engineering disciplines, furthermore, generates a risk of drifting towards what has been already criticized by thinkers of the technological society such as Ellul (1964) or Illich (1982), and referred to as "second order problems" in the sustainability debate. That is, strictly technical solutions to sociotechnical problems serve to increase technicisation and generate new socio-technological problems in a headlong rush, serving ultimately to worsen the situation that is supposed to be mitigated. One typical example of the result of such processes is the often cited "rebound effect," defined for example by Hertwich (2005) in an industrial ecology perspective as "a behavioural or other systemic response to a measure taken to reduce environmental impacts that offsets the effect of the measure." The problem thus lies in the propensity of engineers to develop one-sided technological solutions, or, better said, the general tendency on the part of engineering disciplines to "generate clever solutions for problems that do not exist." Overcoming this problem thus figures hugely in the pursuit of sustainable manufacturing solutions. Specifically, bridges have to be built between disciplines well-rehearsed in asking questions (e.g. humanities) and disciplines adept in developing solutions (e.g. engineering).

Unfortunately, inter- and transdisciplinarity approaches in research remain ridden with obstacles. The major challenges of such approaches are highlighted for example by Schäfer (2013):

- Researchers should be open to broadening their horizons, i.e. acknowledging that collaboration with other disciplines gives them opportunities to address questions that are not accessible within the framework of their own discipline. For example, production technology engineers can develop cleaner production technologies with the help of environmentalists, allowing them to identify the relevant parameters. Empirical observations show that the lack of fulfilment of this basic requirement may be a significant reason for the failure of a large part of transdisciplinary projects.
- Disciplines should acknowledge the epistemic values and methods of other disciplines, which may prove to be particularly thorny between, for example, engineering and humanities—the former being generally based on positivist and the latter on constructivist epistemology.
- Considering that differentiation of technical terminology stands in the way of common understanding between disciplines, the fostering of common understanding requires the development of a common language. This requires in turn that researchers (1) acknowledge terms may have different meanings in their respective disciplines (2) consent to making the effort of identifying potential misunderstandings and defining the terms (3) avoid technical jargon in interdisciplinary exchanges.
- A barrier for openness of researchers towards inter- and transdisciplinarity might lie in the organisation of academia in highly specialized disciplines. In the context of the evaluation of research and allocation of research grants driven by discipline-related quality criteria, inter- and transdisciplinarity research may be disadvantaged.

Although the four difficulties cited here may sound trivial, experiences in major interdisciplinary research projects show that they are decisive indeed. Although convinced by the necessity of developing solutions for sustainability and by the complexity of the problem, researchers may well fail to cultivate interest in interdisciplinarity research and in broadening the focus of their activity. Literature on inter- and transdisciplinary sustainability research already gives some hints on how to address these challenges, that should indeed be more systematically taken into account in the planning and operation of research projects dealing with engineering and sustainability.

4 Conclusions

In this contribution, the current field of research in sustainable manufacturing has been screened, with a particular focus on technology and management. Based on this review, this article provides a definition of the term sustainable manufacturing as well as a structuring framework defining four complementary areas of research: manufacturing technologies ("how things are produced"), product development ("what is being produced"), value creation networks ("in which organisational context") and global manufacturing impacts ("how to make a systemic change"). These layers have been illustrated with examples from current research initiatives addressing analysis, synthesis or transition issues, while their respective principal challenges have been illuminated.

This article emphatically states the equal importance and the complementarity nature of these four layers, at the same time that we likewise underline the necessity of the interdisciplinary nature of action towards sustainable manufacturing. Since individual fields of expertise are unable to grasp the entire complexity of the challenges raised by sustainability, researchers are invited to consider the limits of the solutions they can offer, and to search for broadened perspectives beyond the frontiers of their expertise.

References

Aitamurto, Tanja, Dónal Holland, and Sofia Hussain. 2015. The open paradigm in design research. *Design Issues* 31(4): 17–29. doi:10.1162/DESI_a_00348.

Andersson, Jan Otto, and Mattias Lindroth. 2001. Ecologically unsustainable trade. *Ecological Economics* 37(1): 113–122. doi:10.1016/S0921-8009(00)00272-X.

Arena, Marika, Natalia Duque Ciceri, Sergio Terzi, Irene Bengo, Giovanni Azzone, and Marco Garetti. 2009. A state-of-the-art of industrial sustainability: Definitions, tools and metrics. *International Journal of Product Lifecycle Management* 4(1–3): 207–251. doi:10.1504/IJPLM.2009.031674.

Arnold, Jeanne E., Anthony P. Graesch, Enzo Ragazzini, and Elinor Ochs. 2012. *Life at home in the twenty-first century: 32 families open their doors*, 1st ed. Los Angeles: The Cotsen Institute of Archaeology Press.

Arrow, Kenneth, Bert Bolin, Robert Costanza, Partha Dasgupta, C.S. Carl Folke, Bengt-Owe Jansson Holling, et al. 1995. Economic growth, carrying capacity, and the environment. *Science* 268(5210): 520–521. doi:10.1126/science.268.5210.520.

Bjørn, Anders, and Michael Z. Hauschild. 2013. Absolute versus relative environmental sustainability. *Journal of Industrial Ecology* 17(2): 321–332. doi:10.1111/j.1530-9290.2012.00520.x.

Bjørn, Anders, Manuele Margni, Pierre-Olivier Roy, Cécile Bulle, and Michael Zwicky Hauschild. 2016. A proposal to measure absolute environmental sustainability in life cycle assessment. *Ecological Indicators* 63(April): 1–13. doi:10.1016/j.ecolind.2015.11.046.

Bonvoisin, Jérémy. 2012. *Environmental analysis and ecodesign of information services.* Université de Grenoble.

Bonvoisin, Jérémy, and Jean-François Boujut. 2015. Open design platforms for open source product development: current state and requirements. In *Proceedings of the 20th international conference on engineering design (ICED 15), 8—Innovation and Creativity, 11–22.* Milan, Italy.

Brundtland, Gru, Mansour Khalid, Susanna Agnelli, Sali Al-Athel, Bernard Chidzero, Lamina Fadika, Volker Hauff, et al. 1987. *Our common future.* Oxford University Press.

Buchert, Tom, Alexander Kaluza, Friedrich A. Halstenberg, Kai Lindow, Haygazun Hayka, and Rainer Stark. 2014. Enabling product development engineers to select and combine methods for sustainable design. In *Procedia CIRP, 21st CIRP Conference on Life Cycle Engineering, 15 (January):* 413–18. doi:10.1016/j.procir.2014.06.025.

Costanza, Robert, Ida Kubiszewski, Enrico Giovannini, Hunter Lovins, Jacqueline McGlade, Kate E. Pickett, Kristín Vala Ragnarsdóttir, Debra Roberts, Roberto De Vogli, and Richard Wilkinson. 2014. Development: Time to leave GDP behind. *Nature* 505(7483): 283–285. doi:10.1038/505283a.

Diallo, Yacouba, Alex Etienne, Farhad Mehran, and ILO International Programme on the Elimination of Child Labour. 2013. *Global Child Labour Trends 2008 to 2012.*

Duflou, Joost R., John W. Sutherland, David Dornfeld, Christoph Herrmann, Jack Jeswiet, Sami Kara, Michael Hauschild, and Karel Kellens. 2012. Towards energy and resource efficient manufacturing: a processes and systems approach. *CIRP Annals—Manufacturing Technology* 61(2): 587–609. doi:10.1016/j.cirp.2012.05.002.

Dufrene, Maud, Peggy Zwolinski, and Daniel Brissaud. 2013. An engineering platform to support a practical integrated eco-design methodology. *CIRP Annals—Manufacturing Technology* 62 (1): 131–134. doi:10.1016/j.cirp.2013.03.065.

Edenhofer, Ottmar, R. Pichs-Madruga, Y. Sokona, E. Farahani, S. Kadner, K. Seyboth, A. Adler, et al. (eds.). 2015. *IPCC, 2014: Summary for Policymakers. In Climate Change 2014: Synthesis Report. Contribution of Working Groups I, II and III to the Fifth Assessment Report of the Intergovernmental Panel on Climate Change.* Cambridge, United Kingdom and New York, NY, USA: Cambridge University Press.

Ellul, Jacques. 1964. *The technological society.* Translated by John Wilkinson: Vintage Books.

Energy Saving Trust. 2006. The rise of the machines—a review of energy using products in the home from the 1970s to today.

Ernzer, M., and H. Birkhofer. 2002. Selecting methods for life cycle design based on the needs of a company. In *DS 30: proceedings of DESIGN 2002, the 7th international design conference, Dubrovnik.*

Eurofound. 2012. *Fifth European working conditions Survey.* Luxembourg: Publications Office of the European Union.

European Commission. 2006. *COM(2006)545—action plan for energy efficiency: Realising the potential.*

European Commission. 2013. *EU energy in figures—statistical pocketbook 2013.*

Finkbeiner, Matthias, Erwin M. Schau, Annekatrin Lehmann, and Marzia Traverso. 2010. Towards life cycle sustainability assessment. *Sustainability* 2(10): 3309–3322. doi:10.3390/su2103309.

Garetti, Marco, and Marco Taisch. 2012. Sustainable manufacturing: Trends and research challenges. *Production Planning & Control* 23(2–3): 83–104. doi:10.1080/09537287.2011.591619.

Gutowski, Timothy G., Sahil Sahni, Avid Boustani, and Stephen C. Graves. 2011. Remanufacturing and energy savings. *Environmental Science and Technology* 45(10): 4540–4547. doi:10.1021/es102598b.

Haapala, Karl R., Fu Zhao, Jaime Camelio, John W. Sutherland, Steven J. Skerlos, David A. Dornfeld, I.S. Jawahir, Andres F. Clarens, and Jeremy L. Rickli. 2013. A review of engineering research in sustainable manufacturing. *Journal of Manufacturing Science and Engineering* 135 (4): 041013. doi:10.1115/1.4024040.

Herrmann, Christoph, and Sebastian Thiede. 2009. Process chain simulation to foster energy efficiency in manufacturing. *CIRP Journal of Manufacturing Science and Technology, Life Cycle Engineering* 1(4): 221–229. doi:10.1016/j.cirpj.2009.06.005.

Hertwich, Edgar G. 2005. Consumption and the rebound effect: An industrial ecology perspective. *Journal of Industrial Ecology* 9(1–2): 85–98. doi:10.1162/1088198054084635.

Illich, Ivan. 1982. *Medical nemesis: The expropriation of health.* New York: Pantheon.

International Labour Organization. 2014. *Global employment trends 2014—risk of a jobless recovery?.* Geneva: International Labour Office.

Jayal, A.D., F. Badurdeen, O.W. Dillon Jr., and I.S. Jawahir. 2010. Sustainable manufacturing: Modeling and optimization challenges at the product, process and system levels. Sustainable development of manufacturing systems. *CIRP Journal of Manufacturing Science and Technology* 2(3): 144–152. doi:10.1016/j.cirpj.2010.03.006.

Kara, S., G. Bogdanski, and W. Li. 2011. Electricity metering and monitoring in manufacturing systems. In *Glocalized solutions for sustainability in manufacturing,* ed. Jürgen Hesselbach and Christoph Herrmann, 1–10. Springer, Berlin. http://link.springer.com/chapter/10.1007/978-3-642-19692-8_1.

Kara, S., and W. Li. 2011. Unit process energy consumption models for material removal processes. *CIRP Annals—Manufacturing Technology* 60(1): 37–40. doi:10.1016/j.cirp.2011.03.018.

Kianinejad, K., S. Thom, S. Kushwaha, and E. Uhlmann. 2016. Add-on error compensation unit as sustainable solution for outdated milling machines. *Procedia CIRP* 40(January): 174–178. doi:10.1016/j.procir.2016.01.094.

Knight, Paul, and James O. Jenkins. 2009. Adopting and applying eco-design techniques: A practitioners perspective. *Journal of Cleaner Production* 17(5): 549–558. doi:10.1016/j.jclepro.2008.10.002.

Krüger, Jörg, and The Duy Nguyen. 2015. Automated vision-based live ergonomics analysis in assembly operations. *CIRP Annals—Manufacturing Technology* 64(1): 9–12. doi:10.1016/j.cirp.2015.04.046.

Lapillonne, Bruno, Karine Pollier, and Nehir Samci. 2013. Energy efficiency trends in the EU—lessons from the ODYSSEE MURE Project.

Low, Kay Soon, W.N.N. Win, and Meng Joo Er. 2005. Wireless sensor networks for industrial environments. In *International conference on computational intelligence for modelling, control and automation, 2005 and international conference on intelligent agents, web technologies and internet commerce,* 2:271–76. doi:10.1109/CIMCA.2005.1631480.

Maddison, Angus. 2006. *The world economy.* Paris: Organisation for Economic Co-operation and Development. http://www.oecd-ilibrary.org/content/book/9789264022621-en.

Manzini, Ezio, and François Jégou. 2003. Sustainable everyday. *Design Philosophy Papers,* no. 4.

Meadows, Donella H., Dennis L. Meadows, Jorgen Randers, and William W. Behrens. 1972. *Limits to growth.* Universe Books.

Meadows, Donella H., Jorgen Randers, and Dennis L. Meadows. 2004. *Limits to growth: The 30-year update,* 3rd ed. White River Junction, Vt: Chelsea Green Publishing.

Mihelcic, James R., John C. Crittenden, Mitchell J. Small, David R. Shonnard, David R. Hokanson, Qiong Zhang, Hui Chen, et al. 2003. Sustainability science and engineering: The emergence of a new metadiscipline. *Environmental Science & Technology* 37(23): 5314–5324. doi:10.1021/es034605h.

OECD.Stat. 2016. Gross domestic product (GDP)—Table B1G: gross value added at basic prices, total activity.

Pigosso, Daniela C.A., Henrique Rozenfeld, and Tim C. McAloone. 2013. Ecodesign maturity model: a management framework to support ecodesign implementation into manufacturing companies. *Journal of Cleaner Production* 59(November): 160–173. doi:10.1016/j.jclepro.2013.06.040.

Pigosso, Daniela Cristina Antelmi. 2012. Ecodesign maturity model: A framework to support companies in the selection and implementation of ecodesign practices. Text, Universidade de São Paulo. http://www.teses.usp.br/teses/disponiveis/18/18156/tde-10082012-105525/.

Radebach, Alexander, Hauke Schult, and Jan Christoph Steckel. 2014. On the importance of manufacturing sectors for sustainable development and growth. Lisbon, Portugal.

Rockström, Johan, Will Steffen, Kevin Noone, F. Åsa Persson, Stuart Chapin, Eric F. Lambin, Timothy M. Lenton, et al. 2009. A safe operating space for humanity. *Nature* 461(7263): 472–475. doi:10.1038/461472a.

Roeder, Ina, Matthias Scheibleger, and Rainer Stark. 2016. How to make people make a change—using social labelling for raising awareness on sustainable manufacturing. In *Procedia CIRP, 13th global conference on sustainable manufacturing – decoupling growth from resource use*, 40: 359–64. doi:10.1016/j.procir.2016.01.065.

Rosen, Marc A., and Hossam A. Kishawy. 2012. Sustainable manufacturing and design: Concepts, practices and needs. *Sustainability* 4(2): 154–174. doi:10.3390/su4020154.

Schäfer, Prof Dr Martina. 2013. Inter- und transdisziplinäre Nachhaltigkeitsforschung – Innovation durch Integration? In *Soziale Innovation und Nachhaltigkeit*, ed. Jana Rückert-John, 171–94. Innovation und Gesellschaft. Springer Fachmedien Wiesbaden. http://link.springer.com/chapter/10.1007/978-3-531-18974-1_10.

Seliger, Günther, Carsten Reise, and Pinar Bilge. 2011. Curriculum design for sustainable engineering-experiences from the international master program. In *Advances in sustainable manufacturing*, ed. Günther Seliger, Marwan M. K. Khraisheh, and I. S. Jawahir. Berlin, Heidelberg.

Sproesser, Gunther, Ya-Ju Chang, Andreas Pittner, Matthias Finkbeiner, and Michael Rethmeier. 2015. Life cycle assessment of welding technologies for thick metal plate welds. *Journal of Cleaner Production* 108, Part A: 46–53. doi:10.1016/j.jclepro.2015.06.121.

Sproesser, Gunther, Andreas Pittner, and Michael Rethmeier. 2016. Increasing performance and energy efficiency of gas metal arc welding by a high power tandem process. In *Procedia CIRP, 13th global conference on sustainable manufacturing—decoupling growth from resource use*, 40: 643–48. doi:10.1016/j.procir.2016.01.148.

Stark, Rainer, and Anne Pförtner. 2015. Integrating ontology into PLM-tools to improve sustainable product development. *CIRP Annals—Manufacturing Technology* 64(1): 157–160. doi:10.1016/j.cirp.2015.04.018.

Steckel, Jan Christoph, Ottmar Edenhofer, and Michael Jakob. 2015. What drives the renaissance of coal? In *Proceedings of the National Academy of Sciences (PNAS)* forthcoming.

Stock, Paul, and Rob J.F. Burton. 2011. Defining terms for integrated (multi-inter-trans-disciplinary) sustainability research. *Sustainability* 3(8): 1090–1113. doi:10.3390/su3081090.

Sutherland, John W., Justin S. Richter, Margot J. Hutchins, David Dornfeld, Rachel Dzombak, Jennifer Mangold, Stefanie Robinson, et al. 2016. The role of manufacturing in affecting the social dimension of sustainability. *CIRP Annals—Manufacturing Technology* 65(2): 689–712. doi:10.1016/j.cirp.2016.05.003.

Uhlmann, E., E. Fries, P. Fürstmann, and et al. 2012. Tool wear behaviour of internally cooled tools at different cooling liquid temperatures, 21–27. Istambul, Turkey.

US Central Intelligence Agency. 2016. The World Factbook 2016.

Vezzoli, Carlo, Fabrizio Ceschin, Jan Carel Diehl, and Cindy Kohtala. 2015. New design challenges to widely implement 'sustainable product–service systems.'" *Journal of Cleaner Production*, Special Volume: Why have "Sustainable Product-Service Systems" not been widely implemented? 97(June): 1–12. doi:10.1016/j.jclepro.2015.02.061.

Vitousek, Peter M., Harold A. Mooney, Jane Lubchenco, and Jerry M. Melillo. 1997. Human domination of earth's ecosystems. *Science* 277(5325): 494–499. doi:10.1126/science.277.5325.494.

Ward, Peter T., Alexander Radebach, Ingmar Vierhaus, A. Fügenschuh, and Jan Christoph
    Steckel. 2015. How existing technologies can contribute to reduce global emissions. Helsinki.
Weizsacker, Ernst U. von. 1998. *Factor four: Doubling wealth, halving resource use—a report to
    the Club of Rome*. Earthscan.
WHO. 2016. Occupational Health. *World Health Organisation—Regional Office for Europe*. April
    12. http://www.euro.who.int/en/health-topics/environment-and-health/occupational-health.
World Health Organization. 2013. WHO Global Plan of Action on Workers' Health (2008-2017):
    Baseline for implementation—global country survey 2008/2009—executive summary and
    survey findings.

Sustainability Dynamics

Rainer Stark and Kai Lindow

Abstract Value creation ensures societal prosperity. At the same time, *Sustainable Development* determines the future of global human wellbeing. Both aspects are based on profound environmental, social and economic mechanisms—and both aspects are closely linked. The *Sustainability Dynamics Model* describes the direct and indirect effects of value creation together with the three dimensions of *Sustainable Development*. This contribution introduces and defines the *Sustainability Dynamics Model*. The effects and dynamics are exemplarily shown. Eventually, the link to circular economy is drawn. In the future, the *Sustainability Dynamics Model* can be used as a control model in order to predict consequences of value creation towards environmental, social and economic sustainability.

Keywords Sustainability dynamics model · Sustainable development · Circular economy · Value creation · Consumption and production

1 Dynamics in Value Creation and Sustainable Development

Value creation is a key element for ensuring societal prosperity. In the classic sense, value creation is equated with industrial production to meet the needs of society (cp. Fry et al. 1994). Likewise, *Sustainable Development* (cp. WCED 1987) determines the future of global human wellbeing. In the year 2015, the *United Nations* defined *Sustainable Development Goals* (UN 2015) and targeted them to the year 2030 (Fig. 1).

R. Stark (✉)
Chair of Industrial Information Technology, Institute for Machine-tools
and Factory Management, Technische Universität Berlin, Berlin, Germany
e-mail: rainer.stark@tu-berlin.de

K. Lindow
Technische Universität Berlin, Berlin, Germany
e-mail: kai.lindow@tu-berlin.de

R. Stark et al. (eds.), *Sustainable Manufacturing*, Sustainable Production,
Life Cycle Engineering and Management, DOI 10.1007/978-3-319-48514-0_2

Fig. 1 The United Nations Sustainable Development Goals at a glance (UN 2015, *image source* http://www.un.org/sustainabledevelopment/sustainable-development-goals)

On the 25th of September 2015, 193 countries of the *United Nations General Assembly* adopted a set of sustainable development goals to end poverty, protect the planet and ensure prosperity for all as part of a new sustainable development agenda. Following the adoption, *United Nations* agencies have supported a follow-up campaign on the part of several independent entities, among them corporate institutions and international organizations. The campaign, known as Project Everyone, introduced the term global goals. Its intention is to help communicate the agreed upon *Sustainable Development Goals* to a wider constituency.

For the first time, sustainable consumption and production patterns are specifically mentioned among the seventeen goals (Fig. 2, UN 2015).

The particular claim of goal 12 is to reach out for sustainable consumption and production at "doing more and better with less." The scope ranges from macro- to microeconomic level, from society to individuals, from degradation to pollution along the whole lifecycle, while likewise increasing quality of life. It involves different stakeholders, including business, consumers, policy makers, researchers and retailers among others. Furthermore, this goal requires a systemic approach and cooperation among stakeholders in the entire supply chain, from producer to final consumer.

Ueda et al. take up the basic idea and elaborate: "in association with globalization and networking, every industry in this century is strongly required to contribute to sustainable development, but no solution can be obtained easily when considering the complexity and instability of the social systems. Additionally, maintaining sustainability often creates a dilemma between values of a whole society and values of individuals [...]. Therefore, to resolve this problem, more attention must be devoted to value creation mechanisms" (Ueda et al. 2009).

In this context, both aspects of value creation and sustainable development need to be combined to form *Sustainable Value Creation*. The mechanisms included in value creation and *Sustainable Development* are highly dynamic.

Goal 12. Ensure sustainable consumption and production patterns

12.1 Implement the 10-Year Framework of Programmes on Sustainable Consumption and Production Patterns, all countries taking action, with developed countries taking the lead, taking into account the development and capabilities of developing countries

12.2 By 2030, achieve the sustainable management and efficient use of natural resources

12.3 By 2030, halve per capita global food waste at the retail and consumer levels and reduce food losses along production and supply chains, including post-harvest losses

12.4 By 2020, achieve the environmentally sound management of chemicals and all wastes throughout their life cycle, in accordance with agreed international frameworks, and significantly reduce their release to air, water and soil in order to minimize their adverse impacts on human health and the environment

12.5 By 2030, substantially reduce waste generation through prevention, reduction, recycling and reuse

12.6 Encourage companies, especially large and transnational companies, to adopt sustainable practices and to integrate sustainability information into their reporting cycle

12.7 Promote public procurement practices that are sustainable, in accordance with national policies and priorities

12.8 By 2030, ensure that people everywhere have the relevant information and awareness for sustainable development and lifestyles in harmony with nature

---

Fig. 2 Goal 12 "Ensure sustainable consumption and production patterns" among the seventeen *United Nations Sustainable Development Goals* (UN 2015)

Firstly, value creation is characterised by flows of information, resources, capital and labour among production systems. Secondly, these flows are realized within socio-economic, natural and sociotechnical systems. Thirdly, value creation runs over two major levels:

(a) The micro-economic level manages e.g. value creation in supply-chain of a product and value creation along the lifecycle of a product) and
(b) The macro-economic level manages value creation of an entire branch and value creation among countries and within regions.

The interaction and interdependencies of *Sustainable Value Creation*, therefore, lead to a high dynamic among the different systems and their linkages. Value creation activities and services follow three types of interactions as direct and indirect effects between the three major dimensions of sustainability (environment, society and economy):

1. Causal relations,
2. Magnitude and scale drivers and
3. Latency and timely duration dependencies.

   Causal relations describe the determined effects between a solution and its direct and indirect impact on the three dimensions of sustainability (e.g. a new manufacturing solution and its direct impact on the economy, as well as its indirect impacts on society and environment). The direct and indirect impact is determined by the magnitude and scale of a solution's dissemination (e.g. the societal and environmental impact of a solution becomes measurable due to its increasing market share). The effects and impacts have different latencies and time durations (e.g. the societal and environmental impacts of an established solution have a delay and last a certain period of time). The evaluation and description of these dynamic effects is a scientific task and its solution has to however be practical at the same time.

2   Sustainability Dynamics Model

The *Sustainability Dynamics Model* (SDM) is an instrument for describing the direct and indirect effects of value creation solutions on the three dimensions of sustainability and vice versa. Since value creation solutions are the key elements they become the central focus of the model. The three dimensions of *Sustainable Development* (environment, society and economy) actually represent systems of their own and evolve around the value creation solution (Fig. 3).

   Starting from the value creation solution, direct effects between the solution and each sustainability dimension system can be pinpointed:

- The primary effects on the environment are the use and conversion of energy, materials, greenhouse gases etc.
- The primary effect on the society are the improvement of living standards, the use of products, prosperity etc.
- The primary effects on the economy are manufacturing processes, factories, logistics etc.

   The primary effects on one dimension system can cause impacts on other dimension systems. In addition to causal effects (e.g. between environment and society), the above-mentioned effects in the levels of magnitude and scale as well as latency and time duration can be observed. In this case, the root causes not only primary impacts on one dimension system but also secondary impacts on the other two dimension systems of sustainability. Mutual spiral effects between the sustainability dimension systems can, furthermore, be caused by the intended primary effects.

   The effect of a value creation solution on the dimension systems of sustainability can be defined as an inside-out effect. Even so, cause and effect vary with different

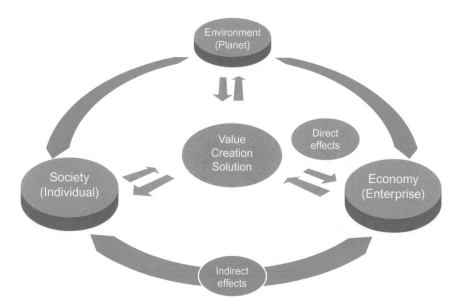

Fig. 3 Introduction of the *Sustainability Dynamics Model* (value creation solution and their direct and indirect effects on sustainable dimension systems)

value creation solutions and their impact on sustainability, among other factors. The *Sustainability Dimension Model* allows an opposite contemplation, which is called an outside-in effect. In this case, the cause can be met in any of the three sustainability dimension systems. This leads to a direct impact on the value creation solution and, additionally, to secondary effects on other sustainability dimension systems through the value creation solution.

An example of outside-in effects is found in the sub-goal 12 "Ensure sustainable consumption and production patterns" of the *United Nations Sustainable Development Goals* (UN 2015). The dynamic effects of sustainable consumption and production play a major role on a macro-economic level, especially in sustainability dimensions and the indirect effects in between. Figure 4 represents the mapping of the eight sub-goals.

In order to illustrate the dynamics in sustainability, the following exemplary goals are revealed:

- 12.2: "By 2030, achieve the sustainable management and efficient use of natural resources" (UN 2015).
- 12.4: "By 2030, achieve the environmentally sound management of chemicals and all wastes throughout their life cycle, in accordance with agreed international frameworks, and significantly reduce their release to air, water and soil in order to minimize their adverse impacts on human health and the environment" (UN 2015).
- 12.5: "By 2030, substantially reduce waste generation through prevention, reduction, recycling and reuse" (UN 2015).

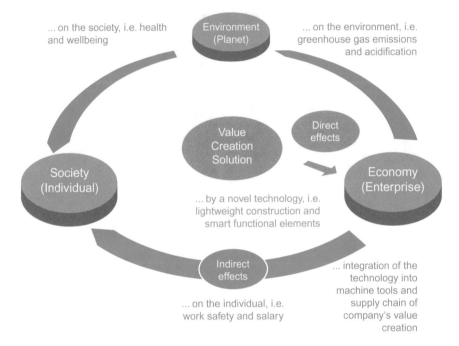

Fig. 4 Mapping of the eight sub-goals of goal 12 "Ensure sustainable production and consumption patterns" of the *United Nations Sustainable Development Goals* and the *Sustainability Dynamics Model*

Goal 12.2 focuses on the overall introduction of principles of sustainable development into country policies and programmes. Regarding the *Sustainability Dynamics Model*, its primary effect lies in the social dimension system. Actions from this dimension system have a direct causal effect on the economic dimension system. Companies within this dimension system have to fulfil sustainable policies and programme demands. That way, the indirect effect is on value creation solutions. Sustainable solutions have to be researched and they need to be applied in manufacturing companies. Depending on the magnitude and scale, an indirect effect on the environmental dimension system takes place and, in return, on the social dimension system on top of that.

Goal 12.4 directly affects the environmental dimension system. The environmentally sound management of chemicals and all wastes throughout their life cycle, along with the significantly reduction of their releases into air, water and soil, together impact both the social and the environmental dimension system at the same time. In order to minimise their impacts on human health, sustainable value creation solutions have to be implemented on a large scale and level of magnitude. These effects occur in latency and timely duration dependencies.

Goal 12.5 deals with the generation and management of waste on a micro- and macroeconomic level. It directly affects the economic dimension system. Technologies and techniques from sustainable value solutions should be applied and used in order to reduce and manage waste from industry. At the same time, products and services that are offered in this dimension system have a causal relationship with its use within the social and individual dimension system. That way, prevention, reduction, recycling and reuse solutions all have an effect on the environmental dimension system.

3   Instantiation of the Model

The direct and indirect effects along the *Sustainability Dynamics Model* can be defined as inside-out effects and outside-in effects. These effects can be either observed when the model is read from the inside (sustainable value creation) to the outside (sustainable dimension systems,) or, vice versa, from the outside the inside. In the following, these two principles are illustrated with two examples.

The first example deals with a novel sustainable manufacturing solution which is based on an innovative manufacturing technology (Fig. 5). This could be gained i.e. by lightweight construction, smart functional elements, improved working accuracy or smart interfaces. The effect on the environmental and social dimension system of the new technology itself is not yet provided. That is, a causal relationship with society and environment can only be found indirectly. However, a direct causal relationship, and in that respect, a direct effect of the new technology, are offered to the economic dimension system. The new technology has to be implemented into a machine tool and into the supply chain. This entails that, a company integrates the solution into their value creation process. Over time, the new solution is in use, indirect effects on the social and environmental dimension system can be found. On the one hand, individuals who are in charge of the new solution are affected i.e. by work safety and salary. Depending on the magnitude and scale of the new solution, the degree of impact on the environmental dimension system is defined, i.e. greenhouse gases and acidification. Furthermore, not only the individual but the whole society is indirectly affected by the environmental impact, i.e. in terms of health and well-being.

The second example deals with the growing awareness among society about sustainable products and services (Fig. 6). In this case, society and individuals demand sustainable solutions. The direct effect is the need for a sustainable value creation solution which can be either a product, a service or a *Product-Service System*. The solution should provide sustainable principles to the customer, i.e. in terms of emissions, noise, safety, costs, recyclability. The indirect effect is basically on manufacturing companies which develop, manufacture and provide the new

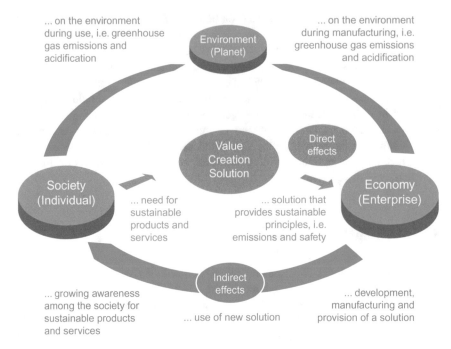

Fig. 5 Inside-out effects of an innovative value creation solution on the economic, social and environmental dimension system

solution to the individual. Depending on the magnitude and scale of the new solution, the impact on the environmental and social dimension system varies, i.e. in terms of greenhouse gas emissions and acidification by manufacturing and use of the solution.

4   Conclusion

The notion of *Sustainability Dynamics* is a new scientific approach which describes the interconnectivity between core dimensions of sustainability and their related internal systems with the system of value creation solutions. The new approach is described within this contribution as a first foundation causal model in pursuit of providing a new basis for describing cross-system sustainability behaviours and influences.

The authors have concentrated on demonstrating the principle power of the model with the help of allocating the sub-targets of goal 12 of the seventeen United Nations Sustainable Development Goals into the causal network of the *Sustainability Dynamics Model*. This goal 12 represents the only goal amongst the seventeen goals which directly addresses sustainable consumption and production patterns critical for sustainable value creation and manufacturing contributions.

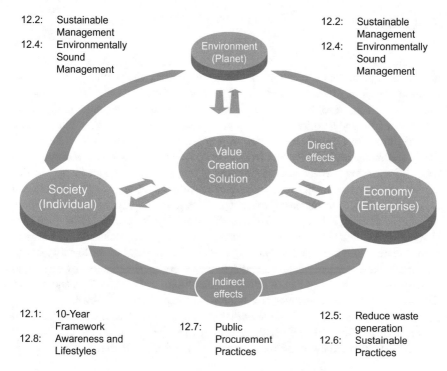

12.2: Sustainable Management
12.4: Environmentally Sound Management

12.2: Sustainable Management
12.4: Environmentally Sound Management

Environment (Planet)

Value Creation Solution

Direct effects

Society (Individual)

Economy (Enterprise)

Indirect effects

12.1: 10-Year Framework
12.8: Awareness and Lifestyles

12.7: Public Procurement Practices

12.5: Reduce waste generation
12.6: Sustainable Practices

Fig. 6 Outside-in effects of a growing awareness among the society for sustainable products and services on the value creation solution and on the economic, social and environmental dimension system

The *Sustainability Dynamics Model* for the first time ever enables the visual and qualitative capabilities for showing the interdependencies and causal effects of value creation solutions (e.g. as part of sustainable product development and sustainable manufacturing) with the major systems of the three sustainability dimensions of environment (planet/earth), economy (enterprises) and society (individual). At this point, in time the *Sustainability Dynamics Model* exists at a foundational level in order to allow high level and principle trade-off discussions and qualitative reasoning.

The next level of the *Sustainability Dynamics Model* is targeted at fostering and expanding the "dynamic" dimension. That is, principles of the model theory *system dynamics* (cp. Sterman 2000) will be utilised in pursuit of quantitative prediction capability. From a knowledge and model depth point of view it will be scrutinized which type of model laws can be integrated robustly. At this point in time it is the authors' belief that the *Sustainability Dynamics Model* bears significant capability to deploy both rule-based dynamic mechanisms as well as big/smart data plug-ins, for the purpose of delivering an increasing level of consequence prediction capability for the contributions of value creation solution towards "measurable" sustainability.

5   Outlook

The major element to transform manufacturing towards "higher sustainability" with respect to global value creation is "resource productivity within a compatible environment" (cp. Bleischwitz et al. 2009). Such target state requires continuous improvements in resource discovery. At the same time, resource productivity remains hugely underexploited as a source of wealth, competitiveness and renewal.

The European Commission started to propose a circular economy strategy (EC 2015) and many business leaders have indeed embraced the circular economy as a path to increasing growth and profitability (Lovins and Braungart 2014). In this manner, the circular economy is gaining increasing attention and offers a potential way for the society to increase prosperity, while reducing dependency on primary materials and energy. In this context, the *Sustainability Dynamics Model* even now at its infancy stage serves as an enabler for explaining basic connections between value creation and circular economy against the background of sustainable development. Furthermore, correlations and coherences could be explained by direct and indirect effects in terms of causal relations, magnitude and scale drivers and latency and time duration dependencies at a micro- and macroeconomic level.

Future expansions of the *Sustainability Dynamics Model,* as depicted in Sect. 4 of this contribution, will deliver the potential to serve as one of the core control models of value creation contributions within the circular economy of the future.

References

Bleischwitz, R., Z. Zhang, and P. Welfens. 2009. *Sustainable growth and resource productivity: Economic and global policy issues.* Greenleaf Publishing.
European Commission (EC). 2015. *Closing the loop—An EU action plan for the Circular Economy.* Brussels, 02.12.2015, COM(2015) 614 final.
Fry, T.D., D.C. Steele, and B.A. Saladin. 1994. A service-oriented manufacturing strategy. *International Journal of Operations & Production Management* 14(10): 17–29.
Lovins, A., and M. Braungart. 2014. *A new dynamic—effective business in a circular economy.* Ellen MacArthur Foundation.
Sterman, J. 2000. *Business dynamics: Systems thinking and modeling for a complex world.* McGraw-Hill Higher Education.
Ueda, K., T. Takenaka, J. Vancza, and L. Monostori. 2009. Value creation and decision-making in sustainable society. *Annals of the CIRP* 58/2: 2009.
United Nations (UN). 2015. *Transforming our world: The 2030 Agenda for Sustainable Development.* United Nations Resolution A/RES/70/1 adopted by the General Assembly on 25 September 2015, New York, 2015.
World Commission on Environment and Development (WCED). 1987. *Our common future.* United Nations General Assembly, Nairobi, 1987.

Enabling Low-Carbon Development in Poor Countries

Jan Christoph Steckel, Gregor Schwerhoff and Ottmar Edenhofer

Abstract The challenges associated with achieving sustainable development goals and stabilizing the world's climate cannot be solved without significant efforts by developing and newly-emerging countries. With respect to climate change mitigation, the main challenge for developing countries lies in avoiding future emissions and lock-ins into emission-intensive technologies, rather than reducing today's emissions. While first best policy instruments like carbon prices could prevent increasing carbonization, those policies are often rejected by developing countries out of a concern for negative repercussions on development and long-term growth. In addition, policy environments in developing countries impose particular challenges for regulatory policy aiming to incentivize climate change mitigation and sustainable development. This chapter first discusses how climate policy could potentially interact with sustainable development and economic growth. It focuses, in particular, on the role of industrial sector development. The chapter then continues by discussing how effective policy could be designed, specifically taking developing country circumstances into account.

Keywords Developing countries · Climate policy · Sustainable development

J.C. Steckel (✉) · G. Schwerhoff · O. Edenhofer
Mercator Research Institute on Global Commons and Climate Change (MCC), Berlin, Germany
e-mail: steckel@mcc-berlin.net

J.C. Steckel · O. Edenhofer
Technische Universität Berlin, Berlin, Germany

J.C. Steckel · O. Edenhofer
Potsdam Institute for Climate Impact Research, Potsdam, Germany

© The Author(s) 2017
R. Stark et al. (eds.), *Sustainable Manufacturing*, Sustainable Production,
Life Cycle Engineering and Management, DOI 10.1007/978-3-319-48514-0_3

1 Introduction

Economic development and poverty eradication (as aimed at in the Sustainable Development Goals, SDGs) have in the past gone hand in hand with the large-scale carbonization of countries' energy systems. That is, countries that have been successful in lifting people out of poverty have also dramatically increased their per-capita emissions, hence contributing significantly to climate change. This trend has recently accelerated by a global renaissance of emission-intensive coal. This renewed embrace of coal is mainly driven by countries that currently have low income, but whose economies are growing rapidly. They are investing in cheap and widely available coal to fuel their increasing energy demand and ongoing industrialization (Steckel et al. 2015). Coal-fired power plants that are currently under construction or planned would—if realized—consume one third (240 Gt of CO_2) of the carbon budget still available to achieve a 2 °C goal (roughly 800 Gt CO_2) (Edenhofer et al. 2016). Six developing or newly industrializing countries (China, India, Vietnam, South Africa, Turkey and Indonesia) are responsible for 85 % of ongoing and planned coal investments. In those countries, the relative prices of coal are usually low despite recent cost reductions of low carbon alternatives, including natural gas and renewable energy (Edenhofer et al. 2016).

Against this background, it comes as no surprise that in order to achieve ambitious climate change mitigation targets, more than half of global mitigation (compared to "business as usual" scenarios based on historic correlations between GDP and carbon emissions) will need to take place in today's low and middle-income countries (Jakob and Steckel 2014). In other words, for the Paris Agreement to be successful, these countries cannot replicate the emission- and energy-intense development pathways of the past, but will need to decouple growing GDP and greenhouse gas emissions. Providing energy by means of low carbon technologies, like renewable energy, biomass, nuclear or fossil fuels in combination with carbon-capture and storage (CCS) is thus one important element in the process of detaching emissions from economic growth (IPCC 2014).

Another way of reducing emissions entails reducing energy use, particularly in the manufacturing sectors. Today, technological differences across economic sectors (i.e. value added per energy input in specific sectors, e.g. the automobile sector) the world over can be multiple orders of magnitude, with poor countries usually employing outdated, inefficient technologies (Kim and Kim 2012). Figure 1 shows that sectoral energy intensity levels in rich countries (listed in Annex I to the UNFCCC) are usually much lower than in developing and newly industrializing countries (non-Annex I countries), with some manufacturing sectors showing differences by multiple orders of magnitude.

Ward et al. (2016) show that equalizing existing differences at least to some extent using technology available today, carries potential for global greenhouse gas (GHG) reductions in the energy sector of 10 Gt CO_2 or more. This result is obtained considering higher order effects—that is, considering the effect of changes in technology on the entire supply chain (first order effects, in contrast, only take

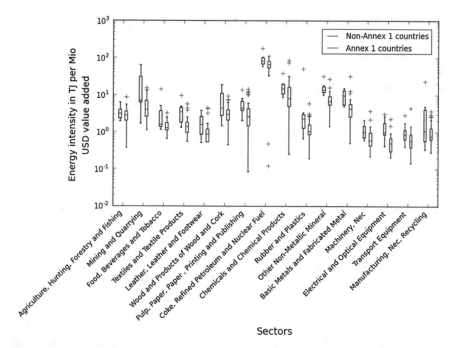

Fig. 1 Distribution of energy intensity of industrial sectors across the World Input-Output Database's (WIOD) regions. *Boxes* represent 25th–75th percentile, *red line* refers to median. *Whiskers* in each direction correspond to 1.5 times the interquartile range. *Black boxplots* represent non-Annex I regions of the UNFCCC, *blue boxplots* corresponds to Annex I regions. *Crosses* represent outliers. *Source* Ward et al. (2016)

direct suppliers into account while multiple layers of the supply chain are ignored). Equalizing existing differences and significantly enhancing energy efficiency levels furthermore likewise play an important role in global mitigation scenarios (IPCC 2014; Luderer et al. 2012).

From an economic point of view, an important question lies in how techno-logical improvements focusing on both the demand side and the investment in low carbon energy systems on the supply side can be incentivized. In this paper we will argue that it is of particular importance to come up with such reward systems that can work in developing country frameworks. Broad agreement among economists holds that a carbon price is the most efficient ("first best") policy instrument. In developing countries, however, carbon prices are hardly ever instituted due to distributive concerns—that is, concerns that the effect of the prices will be dis-tributed unequally amongst the population. A major distributive concern is that carbon prices have a regressive effect, wherein the poor pay proportionally more than the wealthy. Second, there is a concern that carbon prices interfere with economic growth, structural change, involving a shift in importance among dif-ferent sectors in the economy and industrial development (Jakob and Steckel 2014). While this argument is frequently made by policymakers from developing counties, hardly any evidence exists on how exactly structural change and carbon pricing would actually interact.

In this chapter, we will therefore investigate the role of structural change and industrial development on economic growth. Against this background, we will then examine various policy options in developing countries. We will first look into different conceptual possibilities for carbon pricing, including taxes, subsidy removal policies and emissions trading. Second, we will discuss potential barriers specific for developing country environments. We conclude with options for enabling low carbon development in developing countries.

2 Industrialization, Economic Development and Climate Policy

In order to properly assess future developments and evaluate the impacts of envisaged climate policies for affected countries, it is crucial to have a clear picture of the role of specific economic sectors in the process of economic growth. It is particularly important to appreciate the role of energy industry sectors for development. Yet, whereas mitigation scenarios as reviewed in the IPCC (2014) display a high level of technological detail in the energy sector, they usually abstract from modelling economic sectors at a fine resolution. For this reason, some key stylized facts on energy use are not well captured by current climate scenarios. For instance, there is a clear correlation of GDP and energy use up to a certain threshold (Steckel et al. 2013; Steinberger and Roberts 2010). Compared to levels that are observed today, additional energy is undoubtedly needed for covering subsistence needs (Rao et al. 2014) as well as provision of basic infrastructure services (Steckel et al. 2013, 2015). Furthermore, the share of the industry sector in countries' energy demand increases dramatically in development processes before it eventually declines again (Schäfer 2005).

Today, most integrated assessment models (IAMs) that are assessed for the IPCC (2014) and thus constitute the backbone of analyses regarding climate change mitigation, rely on economic models which abstract from differences between sectors. These models however do not take any particular income levels or different economic structures explicitly into account. Instead, they assume that the production factors of labour, capital and (in a subset of models) also energy can be substituted with one another at a given cost. Yet this assumption partly contradicts the empirical observations mentioned above. More realistic modelling of economic growth and associated energy use patterns during industrialization could however indeed substantially affect mitigation costs in developing countries.

Early theories of economic growth focused heavily on the role of specific economic sectors and structural changes. Since the works of Hirschman (1958), the structure of an economy—the composition of economic sectors in the overall economy and how they are interlinked—is commonly conceived of as an important driver for economic growth. Yet as a result of the analytical intractability of such models, one-sector growth models à la Ramsey (1928) and Solow (1956) have become the workhorse models of both economic theory and several IAMs. Structural change has only recently re-emerged as a central topic (Hansen and Prescott 2002), and has been recognized as one of the main factors of future economic growth, in particular in African countries (McMillan et al. 2014).

This recent work shows that during the development process, the forces which drive structural changes are the changing patterns of demand due to increasing incomes and differences in sectoral (labour) productivities. Early in the development process, economies typically have large agricultural sectors and then develop first the industrial and then the service sector (Herrendorf et al. 2014). Convergence of productivities across countries only takes place in manufacturing sectors, or, in countries that have gone through basic structural changes (Rodrik 2013). Countries going through structural changes first diversify their economies (i.e. building up more complex industrial sectors) and then undertake specializing further once they have reached a certain level of affluence (Imbs and Wacziarg 2003).

Recent economic research has probed more deeply into the processes going on within the three major sectors. These authors regard the economy as a network of interconnected products or sectors. In the process of compiling this information into an aggregate index of economic complexity, it turns out that economic complexity (usually measured in the structure of exports) is predictive of economic growth (Hidalgo et al. 2007; Hidalgo and Hausmann 2009) and can even explain economic growth better than aggregated neo-classical growth models (Hausmann 2007; Hausmann and Hidalgo 2011). Some authors (e.g. Hidalgo and Hausmann 2009) moreover presume increasingly complex export structures to be explainable by means of underlying societal capabilities. Increasing complexity is hence related to the increasingly diverse interplay of ingredients that are of general importance for socio-economic development and growth. Radebach et al. (2016) find a clear community structure of economic sectors by using value-added data. Some sectors occupy a central position in the emerging network, mainly light industry sectors, such as textiles and wood products. These sectors can be deemed to be of particular relevance to economic development, as they allow a transition from an agricultural to an industrialized economy. In line with other results from the literature, this result suggests some sectors of being more important for economic growth than others (Fig. 2).

This observation seems to be especially significant considering that underlying capabilities (such as institutions and human capital) relevant for economic growth and development (e.g. Acemoğlu et al. 2005; Acemoğlu and Robinson 2000) depend on increasing complexity. If building up specific (energy- and carbon intensive) sectors enhances spillovers for general economic development and growth, then this indeed yields decisive consequences for climate policy. It follows then, that failing to go through the process of the industrial stage proves detrimental to an economy aiming at economic growth and sustainable development. Yet more central in the pursuit of sustainable development are the factors of innovation and technological development, or, *sustainable manufacturing*.

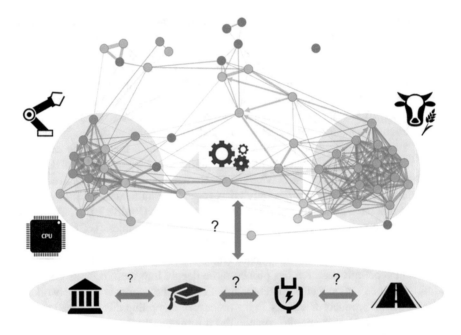

Fig. 2 Stylized representation of the role of manufacturing sectors for structural change and economic development. To the *right* (*green dots*), mainly agricultural sectors can be seen, while high-tech (*dark blue*) and service sectors (*yellow*) sectors are mainly found on the *left hand side*. Certain sectors bridge those communities (light manufacturing sectors, *light blue*), which has given rise to the hypothesis that those sectors are important for building up societal capabilities, including institutions, education, and infrastructure. Adapted from Radebach et al. (2016)

This observation thereby yields important insights for the design of policy instruments in developing countries. Climate policy that discourages investments in manufacturing sectors and decelerates structural change might therefore indeed prove harmful to development—an argument often brought forward by developing countries themselves. For example, from the very onset of the UNFCCC Conference of the Parties in Paris, India's Prime Minister Modi proceeded to highlight, while acknowledging the challenges of climate change for India, that his country will further invest in coal to fuel its energy needs and ensure its right to development.

On that token, the following section will explore the policy options for enhancing low-carbon development in developing countries in more detail.

3 Incentivizing Change—Carbon Pricing in a Developing Country Context

From an economic theory point of view, carbon pricing is the *sine qua non* of climate policy, a precept broadly agreed upon by economists (see e.g. Acemoglu et al. 2012; Stiglitz 2016; Weitzman 2014). A global price on carbon is generally believed to be a key solution for settling the climate problem, which was recently prominently reiterated by MacKay et al. (2015). Carbon prices ensure that negative implications and damages of emissions—including changes in the climate—are readily transparent and therefore taken into account by market participants, and hence incorporated into investment decisions. Applied to the entire economy, they also ensure that loopholes can be avoided and transaction costs can be kept to a minimum. Other policy instruments, like research subsidies and technology standards, furthermore, have proven quite successful in reducing the energy and carbon intensity of the targeted sectors or products, but at the same time do not prevent increasing emissions in other areas of the economy. This is an effect described as the "rebound effect" (Arvesen et al. 2011; Gillingham et al. 2016)

In this context, it is important to keep in mind that carbon prices can be implemented in a wide variety of ways. While the straightforward method is obviously imposing a tax on carbon (which again can be levied at various points of regulation, up- or downstream), a carbon price can also be applied in the form of a quantity-based instrument, i.e. an emissions trading scheme. Following the logic applied in the Kyoto Protocol, it has long been discussed as a viable means of implementing an international carbon market. In such a trading scheme, the amount of total emissions is capped, while emission allowances are allocated to countries. The allocation is often inspired by an ethical principle and results from specific negotiations between countries, e.g. equal emission rights per capita. An international carbon price would then be established on the grounds of supply and demand for emission certificates. Allocation schemes could be designed in such a way that they favour developing countries insofar as they ensure that they are compensated for the potential incremental costs attached to low-carbon technologies.

While countries are increasingly implementing carbon pricing schemes (in particular OECD countries, World Bank 2015), high fossil fuel subsidies have led to a de facto subsidy for carbon (i.e. a negative carbon price) at the global level (Coady et al. 2015). Foregoing fossil fuel subsidies is hence an important first step in incentivizing climate change mitigation, particularly in developing countries.

While affecting the relative price with carbon prices is appealing conceptually, this process runs up against copious obstacles in developing countries, some of which do not exist in this form in developed countries. First, financing costs are usually higher in developing countries. Typically, interest rates are higher and access to capital is more difficult than in developed countries, as are the political and regulatory risks incurred by investors. Both factors lead to weighted average costs of capital in developing countries being significantly higher than in OECD countries (Schmidt 2014). In this market environment, raising (or implementing) a price on carbon would increase energy prices, but not necessarily lead to investment in low-carbon and energy efficient technologies. As those are usually more capital intensive than dirty technologies (Schmidt 2014) a price on carbon can be ineffective in terms of triggering low-carbon investments and hence remains ineffective (Hirth and Steckel, under review). Additional policy instruments designed to alleviate investor risk and buy down technology costs thus might be needed in addition to carbon pricing in order to incentivize low-carbon development.

Second, a range of economic analyses on carbon pricing implicitly presume a liberalized energy market that allows for price signals to be passed through. However, this stands in stark contrast to the (generally) non-liberalized nature of energy markets in many developing countries (Goldblatt 2010; Wisuttisak 2012). Non-liberalized energy markets however indeed grossly impact the effectiveness of carbon pricing. When the government (or one of its agencies) is directly responsible for energy investments, it therefore has to take on payment of the carbon price itself. Unless government agencies, and in particular the energy utility, are made fully responsible for their individual financial performance, the carbon price is unlikely to have any strong incentive effect. This is in particularly true when the government aims to keep energy prices as low as possible, e.g. to prevent negative income effects on poor households or out of the interest in competiveness. In developing countries, where energy utilities are responsible for a large part of total emissions, this situation can mean that total emissions are hardly affected by the carbon price.

Hence, a third obstacle is rooted in distributive concerns. If carbon prices have the desired incentivizing effect, they inevitably cause higher energy prices. Low energy prices, however, are ostensibly considered to be an essential channel for supporting the poor in many countries and are often subsidized for that very reason. What's more, energy prices are considered to be a critical element in the pursuit of the competitiveness of the country's overall economy in the global marketplace. Indeed rising energy prices frequently lead to public protests and societal unrest. Yet balancing distributive issues of rising energy prices is far from impossible. Foregoing fossil fuel subsidies in Indonesia or Iran, for example, have been complemented by transfer schemes favouring poor households (Lindebjerg et al. 2015).

The quality of institutions is also relevant when considering carbon pricing options in developing countries. One frequently proposed model for implementing international carbon prices is an international carbon market, which considers equity issues by means of allowance allocation schemes that favour developing countries. Jakob et al. (2015) emphasize that related transfers could be in the form of resource rents, for example, that yielded negative implications on long-term growth in the past—often referred to as "resource curse". Under such conditions, developing countries might not be able to absorb the carbon rent in a productive way. Low institutional quality and high rates of corruption might also have proven to be pertinent in cases where a pricing instrument was in place related to administrative efforts at monitoring market participants. Even in the EU ETS, some have reported that information asymmetries between regulators and firms have led to reported cases of fraud (Nield and Pereira 2011).

4   Conclusion: Climate Policy Solutions for Developing Countries

Overall, a price on carbon is seen to have the effect of penalizing carbon emissions. To avoid paying this price, firms could reduce emissions per unit of energy by employing low-carbon technology and reducing energy use by improving energy efficiency. However, specific market environments featuring rather low institutional quality, coupled with rather high inequality, high capital costs and regulated energy markets, to mention only a few factors, need all to be taken into careful consideration when crafting the design of policy instruments. Most importantly, it needs to be acknowledged that distributional concerns, both regarding the poorest parts of countries' populations as well as decelerated economic growth (i.e. slower convergence to developed countries' income levels), likewise figure into the equation in a huge way for policy makers in developing countries.

Given this background, Jakob et al. (2016) propose using revenues from carbon pricing to finance infrastructure investment. In many countries, the revenues which a government can collect from taxing CO_2, can then be utilized to finance SDGs, e.g. access to water, sanitation or electricity. In the case of Nigeria, Dorband (2016) shows that a carbon tax deployed in this way turns out to be largely progressive, and hence can alleviate distributional concerns.

In addition, it will be necessary to institute de-risk measures for investments in low-carbon technologies. While one possibility could be to implement subsidies in addition to carbon prices, it may well be useful to offer additional securities to companies that aim to invest in developing countries. Those securities today are often granted to fossil fuels (Coady et al. 2015).

Moreover, carbon taxes can be levied downstream, for example, at the sale of the final good to the consumer, or upstream, where a fossil fuel is extracted or imported. Given issues with institutional quality, it seems that levying carbon taxes upstream is the most useful mechanism for introducing a carbon tax in developing countries. Even when markets remain regulated in this way, fossil fuel costs increase. This would also be relevant for investment decisions taken by governments themselves.

An open question however remains with regards to carbon taxes possibly ushering in a decelerating force on industrial development. As literature shows possible positive spillovers from a more complex economy, it might therefore be necessary to come up with industrial policy to complement climate policy in order to alleviate negative effects on growth and development. Future research will be needed to better appreciate the precise relationship between industrial development and climate policy.

Finally, the UNFCCC demands common but differentiated responsibility and burden sharing, implying support from developed for developing countries both for climate change adaptation and mitigation. Based on those fundamental principles, international climate finance (e.g. by the Green Climate Fund) is supposed to support the low-carbon transformation of developing countries. While international climate finance is slowly under way, it is still rather unclear how exactly it will be disbursed. It will be an interesting question for future climate negotiations to tackle how to redesign international climate finance to support structural transformations towards low-carbon development and economic leapfrogging in pursuit of climate change mitigation.

References

Acemoglu, Daron, Philippe Aghion, Leonardo Bursztyn, and David Hemous. 2012. The environment and directed technical change. *American Economic Review* 102(1): 131–166. doi:10.1257/aer.102.1.131.

Acemoğlu, Daron, Simon Johnson, and James A. Robinson. 2005. Institutions as a fundamental cause of long-run growth. In *Handbook of economic growth*, ed. Philippe Aghion and Steven Durlauf, 1:385–472. Amsterdam: Elsevier.

Acemoğlu, Daron, and James A. Robinson. 2000. Political losers as a barrier to economic development. *American Economic Review* 90(2): 126–130.

Arvesen, Anders, Ryan M. Bright, and Edgar G. Hertwich. 2011. Considering only first-order effects? How simplifications lead to unrealistic technology optimism in climate change mitigation. *Energy Policy* 39: 7448–7454.

Coady, David, Ian Parry, Louis Sears, and Shang Baoping. 2015. How large are global energy subsidies? WP/15/105. IMF Working Paper. Washington D.C.: International Monetary Fund. https://www.imf.org/external/pubs/ft/wp/2015/wp15105.pdf.

Dorband, Ira. 2016. Using revenues from carbon pricing to close infrastructure access gaps— distributional impacts on Nigerian households. Master thesis, FU Berlin.

Edenhofer, Ottmar, Jan Christoph Steckel, Michael Jakob, and Christoph Bertram. 2016. How cheap coal threatens the paris agreement. MCC Working Paper.

Gillingham, Kenneth, David Rapson, and Gernot Wagner. 2016. The rebound effect and energy efficiency policy. *Review of Environmental Economics and Policy* 10(1): 68–88. doi:10.1093/reep/rev017.

Goldblatt, Michael. 2010. Comparison of emissions trading and carbon taxation in South Africa. *Climate Policy* 10(5): 511–526.

Hansen, G., and E. Prescott. 2002. Malthus to Solow. *American Economic Review* 92: 1205–1217.

Hausmann, Ricardo. 2007. What you export matters. *Journal of Economic Growth* 1: 1–25.

Hausmann, Ricardo, and César Hidalgo. 2011. The network structure of economic output. *Journal of Economic Growth* 16: 309–342. doi:10.1007/s10997-011-9071-4.

Herrendorf, Berthold, Richard Rogerson, and Ákos Valentinyi. 2014. Growth and structural transformation. In *Handbook of economic growth*, ed. Philippe Aghion and Steven Durlauf, 2:855–941. Amsterdam: Elsevier.

Hidalgo, César A., and Ricardo Hausmann. 2009. The building blocks of economic complexity. *Proceedings of the National Academy of Sciences* 106(26): 10570–10575. doi:10.1073/pnas.0900943106.

Hidalgo, Cesar A., Bailey Klinger, Albert-L. Barabási, and Ricardo Hausmann. 2007. The product space conditions the development of nations. *Science* 317(5837): 482–487. doi:10.1126/science.1144581.

Hirschman, A.O. 1958. *The strategy of economic development. Yale studies in economics 10*. New Haven, CT, USA: Yale University Press.

Imbs, Jean, and Romain Wacziarg. 2003. Stages of diversification. *American Economic Review* 93 (1): 63–86. doi:10.1257/000282803321455160.

IPCC. 2014. Climate change 2014: mitigation of climate change. Contribution of Working Group III to the Fifth Assessment Report of the Intergovernmental Panel on Climate Change, ed. Ottmar Edenhofer, Ramón Pichs-Madruga, Youba Sokona, E. Farahani, S. Kadner, K. Seyboth, A. Adler, et al. Cambridge, United Kingdom and New York, NY, USA: Cambridge University Press. http://mitigation2014.org.

Jakob, Michael, Claudine Chen, Sabine Fuss, Annika Marxen, and Ottmar Edenhofer. 2015. Development incentives for fossil fuel subsidy reform. *Nature Climate Change* 5(8): 709–712.

Jakob, Michael, Claudine Chen, Sabine Fuss, Annika Marxen, Narasimha D. Rao, and Ottmar Edenhofer. 2016. Carbon pricing revenues could close infrastructure access gaps. *World Development*. doi:10.1016/j.worlddev.2016.03.001.

Jakob, Michael, and Jan Christoph Steckel. 2014. How climate change mitigation could harm development in poor countries. *WIREs Climate Change* 5: 161–168. doi:10.1002/wcc.260.

Kim, Kyunam, and Yeonbae Kim. 2012. International comparision of industrial CO_2 emission trends and the energy efficiency paradox utilizing production-based decomposition. *Energy Economics* 34(5): 1724–1741.

Lindebjerg, Erik S., Wei Peng, and Stephen Yeboah. 2015. Do policies for phasing out fossil fuel subsidies deliver what they promise? Working Paper 2015-1. UNRISD Working Papers. Geneva, Switzerland: United Nations Research Institute for Social Development (UNRISD). http://www.unrisd.org/80256B3C005BCCF9/httpNetITFramePDF?ReadFormandparentunid= 170D2DA8A96A5352C1257DC40050C975andparentdoctype=paperandnetitpath=80256- B3C005BCCF9/(httpAuxPages)/170D2DA8A96A5352C1257DC40050C975/$file/Linde- bjerg%20et%20al.pdf.

Luderer, Gunnar, Valentina Bosetti, Michael Jakob, Jan Christoph Steckel, Henri Waisman, and Ottmar Edenhofer. 2012. The economics of decarbonizing the energy system—results and insights from the RECIPE model intercomparison. *Climatic Change* 114(1): 9–37.

MacKay, David J.C., Peter Cramton, Axel Ockenfels, and Steven Stoft. 2015. Price carbon—I will if you will. *Nature* 526(7573): 315–316. doi:10.1038/526315a.

McMillan, Margaret, Dani Rodrik, and Íñigo Verduzco-Gallo. 2014. Globalization, structural change, and productivity growth with an update on Africa. *World Development* 63: 11–32. doi:10.1016/j.worlddev.2013.10.012.

Nield, K., and R. Pereira. 2011. Fraud on the European Union emissions trading scheme: Effects, vulnerabilities and regulatory reform. *European Energy and Environmental Law Review* 20(6): 255–289.

Radebach, Alexander, Jan Christoph Steckel, and Hauke Ward. 2016. Patterns of sectoral structural change—empirical evidence from similarity networks. http://ssrn.com/abstract= 2771653.

Ramsey, F.P. 1928. A mathematical theory of saving. *The Economic Journal* 38(152): 543–559.

Rao, Narasimha D., Keywan Riahi, and Arnulf Grubler. 2014. Climate impacts of poverty eradication. *Nature Climate Change* 4(9): 749–751. doi:10.1038/nclimate2340.

Rodrik, Dani. 2013. Unconditional convergence in manufacturing. *Quarterly Journal of Economics* 128(1): 165–204. doi:10.1093/qje/qjs047.

Schäfer, Andreas. 2005. Structural change in energy use. *Energy Policy* 33: 429–437.

Schmidt, Tobias S. 2014. Low-Carbon investment risks and de-risking. *Nature Climate Change* 4 (4): 237–239.

Solow, R. 1956. A Contribution to the Theory of Economic Growth. *The Quarterly Journal of Economics* 70: 65–94.

Steckel, Jan Christoph, Robert J. Brecha, Michael Jakob, Jessica Strefler, and Gunner Luderer. 2013. Development without energy? Assessing future scenarios of energy consumption in developing countries. *Ecological Economics* 90: 53–67. doi:10.1016/j.ecolecon.2013.02.006.

Steckel, Jan Christoph, Ottmar Edenhofer, and Michael Jakob. 2015. Drivers for the renaissance of coal. *Proceedings of the National Academy of Sciences* 112(29): E3775–E3781.

Steinberger, Julia K., and J. Timmons Roberts. 2010. From constraint to sufficiency: The decoupling of energy and carbon for human needs, 1975–2005. *Ecological Economics* 70: 425–433.

Stiglitz, Joseph E. 2016. How to restore equitable and sustainable economic growth in the United States †. *American Economic Review* 106(5): 43–47. doi:10.1257/aer.p20161006.

Ward, Hauke, Alexander Radebach, Ingmar Vierhaus, Armin Fügenschuh, and Jan Christoph Steckel. 2016. Reducing global CO_2 emissions with the technologies we have. MCC Working Paper.

Weitzman, Martin L. 2014. Can negotiating a uniform carbon price help to internalize the global warming externality? *Journal of the Association of Environmental and Resource Economists* 1 (1/2): 29–49. doi:10.1086/676039.

Wisuttisak, Pornchai. 2012. Regulation and competition issues in Thai electricity sector. *Energy Policy* 44: 185–198.

World Bank. 2015. *State and Trends of Carbon Pricing 2015*. Washington, DC: World Bank.

Part II
Solutions for Sustainability-Driven Development of Manufacturing Technologies

Dr.-Ing. Fiona Sammler, Institute for Machine Tools and Factory Management, TU Berlin, Germany

Dr.-Ing. Nils F. Nissen, Fraunhofer Institute for Reliability and Microintegration, Berlin, Germany

At the core of sustainable manufacturing is what happens on the factory floor. While many other aspects—such as linking and planning production steps, connecting different players and production sites around the globe, or analysing the sustainability effects of the choices we make—are essential in moving in the direction of sustainability, we still need improvements and innovations at the level of the individual machine tool and manufacturing process.

The sustainable factory floor is by nature more diverse and complex than traditional manufacturing since it needs to be adaptable according to the geographical location, the skill level of employees, the locally available materials and resources, and the individual workers' wellbeing. This chapter focuses on the development of sustainable machine tools and manufacturing processes for the production of industrial goods, which when scaled up, can impact the design of a factory floor. Of course, the goal of sustainable manufacturing processes is to allow a higher value creation with reduced resource consumption.

Whilst the general focus in the production industry remains on the reduction of costs and the production of high quality goods, a higher priority has been ascribed to the overall environmental, economic and social impact of the production process. However, incorporating the solutions developed into an industrial setting has thus far been limited to individual measures, with various countries responding quite differently to the challenge of sustainable manufacturing all the while. Significant barriers persist in hindering the development and implementation of sustainable manufacturing processes. In particular, barriers related to high costs, a lack of availability of funds for green projects, a lack of support from leaders and a lack of standardisation still represent the status quo. In pursuit of the adoption of sustainability-driven manufacturing solutions thus, a clear, tangible advantage for the production industry has to be introduced. Such an advantage is found in increased consumer interest in the sustainability factor, an advantage on account of government levies or cost savings in the long-term due to better recycling options.

Overall, it is essential that the use of renewable resources for the production of goods be favoured. This is however quite often not fully realisable given the simultaneous necessity of achieving highest quality production standards in series production. When using non-renewable resources, reconfigurability and re-manufacturing must constitute the focus on a much larger scale. The first contribution of this section of the book presents a vision for the future of the machine tool industry in the form of a LEG2O machine tool system. It is a machine tool made of passive and smart building blocks with integrated sensor nodes and can be built according to the demand of a certain product, process, factory, worker skill level or location. The high level of mass customization in today's production industry can therefore be achieved whilst using a defined level of resources without the need for additional investment.

Whilst significant progress has been made in addressing the environ-mental dimension of the triple bottom line, for example, through the use of new, light-weight materials such as carbon fibre reinforced plastics and other composite materials which have found application in the automotive, aeronautical and machine tool industries, it is the social dimension which must likewise consume our full attention in the future. This can take the form of developing technologies which allow organisations, particularly in developing countries, to create a higher level of value creation to strengthen their manufacturing industry and thus their ability to cater to the needs of the next generation. An example of such a technology is discussed in the second contribution of this chapter in the form of the recently developed Accuracy Increasing Add-On System (AIAS).

The final contribution presents a solution for the energy and cost inten-sive cooling of machining processes, a necessary intervention in order to avoid t he overheating of tools and equipment in a large proportion of cutting processes. The development of an internally cooled turning tool, which allows for the cooling of the tool without cooling lubricant contacting the workpiece, opens up new possibilities for saving on cooling lubricant. This in turn increases the ease of chip recycling, the reduction of energy required per part produced, and from a social point of view, the avoidance of skin irritations due to worker contact with the hazardous cooling lubricant.

To summarise, a strong requirement on the part of the production industry lies in developing solutions and acting responsibly with regards to the triple bottom line. Allowing developing countries to build up their own indigenous manufacturing industry is key to combatting poverty and, as such, solutions must be developed to help the constituents of the developing world help themselves. Such local production networks are not to be seen as competition to the import of goods from de-veloped countries. Instead, they form the basis for deepening the relationships. At the same time, the equipment demands in developed countries are changing rapidly, in contrast to the traditional circumstances wherein production equipment changed very gradually over time. The contributions in this chapter give individual examples of such ideas, with the aim of allowing the reader to develop one's own thoughts on how exactly today's manufacturing industry can and must change in the upcoming years.

Sustainable Solutions for Machine Tools

Eckart Uhlmann, Klaus-Dieter Lang, Lukas Prasol, Simon Thom, Bernd Peukert, Stephan Benecke, Eduard Wagner, Fiona Sammler, Sebastian Richarz and Nils F. Nissen

Abstract Environmental, economic and social changes of any significant propor- tions cannot take place without a major shift in the manufacturing sector. In today's manufacturing processes, economic efficiency is realised through high volumes with the use of specialised machine tools. Change in society, such as in the form of mobility and digitisation, requires a complete overhaul in terms of thinking in the manufacturing industry. Moreover, the manufacturing industry contributes over 19 % to the world's greenhouse gas emissions. As a consequence of these issues, a demand for sustainable solutions in the production industry is increasing. In par- ticular, the concept of "cost" in manufacturing processes and thus the "system boundaries" within the production of the future has to be changed. That is, a great number of aspects to the machine tool and production technology industries can be improved upon in order to achieve a more sustainable production environment. Within this chapter, the focus lies on microsystem technology enhanced modular machine tool frames, adaptive mechatronic components, as well as on internally- cooled cutting tools. An innovative machine tool concept has been developed recently, featuring a modular machine tool frame using microsystem technology for communication within the frame, which allows for a high level of flexibility. Furthermore, add-on upgrading systems for outdated machine tools—which are particularly relevant for developing and emerging countries—are poised to gain in importance in the upcoming years. The system described here enables the accuracy of outdated machine tools to be increased, thus making these machine tools com- parable to modern machine tool systems. Finally, the cutting process requires solutions for dry machining, as the use of cooling lubricants is environmentally damaging and a significant cost contributor in machining processes. One such solution is the use of internally cooled cutting tools.

E. Uhlmann (✉) · L. Prasol · S. Thom · B. Peukert · F. Sammler · S. Richarz
Institute for Machine Tools and Factory Management, TU Berlin, Berlin, Germany
e-mail: uhlmann@iwf.tu-berlin.de

K.-D. Lang · S. Benecke · E. Wagner
Research Center for Microperipheric Technologies, TU Berlin, Berlin, Germany

N.F. Nissen
Fraunhofer Institute for Reliability and Microintegration, Berlin, Germany

© The Author(s) 2017
R. Stark et al. (eds.), *Sustainable Manufacturing*, Sustainable Production,
Life Cycle Engineering and Management, DOI 10.1007/978-3-319-48514-0_4

Keywords Flexible machine tools · Modular frames · Microsystem technology · Outdated machine tools · Adaptronic add-ons · Dry machining

1 Introduction

The manufacturing industry influences economic, ecological and social development worldwide. Industrial energy consumption has been increasing in most developed and undeveloped countries over the last decades. Nearly a third of the worldwide energy consumption as well as CO_2-emissions are related to the production industry (International Energy Agency 2007). Furthermore, an increasingly mobile and digital society is calling for new customised technical solutions to a diverse array of products both anytime and anywhere.

In pursuit of sustainable change in the manufacturing industry, it is necessary to develop innovative solutions for machine tools as well as for production processes. To meet the demands of the global market on top of that, it's important to identify "new ways" for sustainable solutions for machine tools which may serve to generate a long-term effect in the production industry.

To impact the sustainability of machine tools, machine tool frames, in particular, must constitute a central focus. Machine tools are "static" in general. Current flexible manufacturing systems are able to handle several production situations. As a result, they are rarely fully exploited and usually "over engineered" and therein require a lot of engineering hours and raw materials. They start their product life with a negative environmental burden as a result.

Even flexible manufacturing systems are not suitable for handling the batch size "one" and therefore short development times are required that can be adapted to new requirements. These challenges are the main drivers for the future development of the machine tool industry, taking into account increasingly scarce resources at hand.

The increase in accuracy of machine tool frames, the usage of mechatronics regarding the accuracy of axes, as well as the "sustainable engineering" of solutions and recycling of components and equipment, constitute the main parts of the research work presented in this chapter. By applying modular machine tool frames using microsystem technology, the accuracy of these applications has been increased significantly. Modular machine tool frames can be used again and again and remain automatically up to date. Furthermore, the realisation of the product batch size "one" is possible if required. Through the use of adaptronic components, the accuracy of machine tools is increased, meaning older machine tools can stay abreast of latest developments. These adaptive components can furthermore be used in modular machine tool frames. By applying innovative tools for machining, sustainable solutions for machine tools are thus being identified from several different angles.

2 Technological Concepts

2.1 Microsystems Technology

Microelectronics constitutes the core of the up-and-coming paradigm "Industrie 4.0" and "cyber physical systems," including supply chains and manufacturing environments. Small distributed systems such as wireless sensor nodes (WSN) and -systems (WSS) are mainly applied within safety systems, control systems (closed loop regulatory systems, closed loop supervisory systems, open loop control systems), monitoring systems (alerting systems, information gathering systems), e.g. the monitoring process of production equipment, yet also feature logistics support with electronic functions beyond radio-frequency identification (RFID) (Schischke 2009). The recent progress in the research of WSN application meets the requirements of manufacturing environments such as functional integrity, robustness, miniaturization and low energy consumption, as well as the more general industrial requirements of low cost, interoperability, resistance to noise and co-existence, self-configuration and organisation, scalability, data allocation and processing, resource efficient design, adaptive network optimization, time synchronization, fault tolerance and reliability, application specific design, and secure system design (Zurawski 2009). The use of WSN poses ecological questions that need to be balanced efficiently on the part of the hard- and software designers involved by means of customised WSN architecture and a WSS layout which closely follow the functional application or use-case scenario whilst at the same time maintaining a low environmental footprint.

WSN for industrial environments can be deployed remotely from the actual point of measurement using complex data acquisition techniques for gauging, constrained by environmental noise. Autonomous WSNs are applied physically in the peripheral environment of the wired grid without direct cable access. Hence, these systems enable sensing tasks at frequently changing or remote locations that cannot be accessed by conventional measurement equipment (Ovsthus and Kristensen 2014). Basic WSN **hardware design** features a central processing unit (microcontroller) that is linked to memory, a communication unit (radio frequency (RF) transceiver including an antenna), sensor units, an independent energy source (energy harvester or battery) and, optionally, a human-machine-interface (HMI).

In the case of distributed WSNs, the **energy efficiency** is a critical factor, as they carry limited energy storage. While wired concepts allow the focus to be on the quality of service (QoS), wireless concepts aim to achieve primarily power conservation at the expense of, for example, lower throughput or higher transmission delay. To meet the requirements of energy efficiency while conserving reliable messaging, industrial settings call for an application of specific communication technologies including RF interference problems, more complex circuitry, individual software algorithm design, WSN topology, such as star or multihop meshed communication network and costs. An overview of radio frequencies involved is given in (Rault et al. 2014). Data losses and communication reliability that appear in industrial settings suffer from noise, co-channel interferences and multipath propagation resulting from such typical obstacles as stationary or moving objects (noise) and RF interferences from other devices (Ovsthus and Kristensen 2014).

Wireless **sensor networking** technology in terms of protocols and standards for the so-called industrial internet of things (IIoT) attempt to combat the obstacles mentioned (Hu 2015). Based on the IEEE 802.XX standard, the derivations ISA 100 and wireless HART are used for applications in process automation since they are considered to be energy efficient, robust and reliable. ZigBee (IEEE 15.4) is considered to be geared towards low energy consumption, low cost and security. As a middleware publishing/subscribing protocol, the Message Queuing Telemetry Transport (MQTT) protocol is being considered for reliable messaging due to its lightweight architecture (Sheltami et al. 2015).

With greatly reduced energy consumption, it is also becoming feasible to employ efficient battery and **energy harvesting technology** as decentralised energy sources. Primary batteries offer the highest performance with about 3000 J/cm^3. Manufacturing environments offer high potential for broadband vibration, fluidic or thermoelectric energy harvesting sources at about 40 $\mu W/cm^2$ to 1 mW/cm^2. In the trade-off, high performance computing and high sampling rates stand up against low-power and miniaturised applications (Beeby 2006, Gungor and Hancke 2009; Elvin and Erturk 2013).

A broader investigation of the trade-off between **environmental** benefits and the negative impact of the additional microsystems including wireless sensor nodes (WSNs) was conducted in the German technology assessment study "Innovations- und Technikanalyse Autonomer Verteilter Mikrosysteme" (Autonomous Distributed Sensor Systems) (Schischke 2009). Qualitative results show that primary effects, e.g. resource consumption or recycling, are mainly negative, while indirect impact, such as production efficiency, are positive. Moreover, the long-term compatibility between the different lifecycles and concepts of machine tool components and electronics remain an interesting research topic.

To date, a link between the impact of microsystems on component level and **modularity** leading to further improvement recommendations on the system level remains missing. The question of how to support the designer of electronic systems with easy-to-use indicators while addressing sustainability issues has been addressed by (Wagner et al. 2016). The design methodology developed serves to connect common electronic components to their contained materials and selected impact types like cumulative energy demand or recyclability. From the system assessment standpoint, there is a need for evaluations of the trade-off between more functionality and more resource impact for 25+ years use time.

2.2 Reconfigurable Machine Tools

The design process of machine tools represents a major investment in tangible and intangible resources for machine tool manufacturer and consumer. Lead times are especially high in the case of individualised machine tools and assembly lines. From a sustainability perspective, long lifecycles of machines of over 25 years are difficult to manage, given the volatility of product variants and low batch sizes of today's global market. As the production of machine tools has high relevance in the world economy (Verein Deutscher Werkzeugmaschinenfabriken e.V 2015), machine tool manufacturers are therefore interested in shortening delivery times, increasing flexibility and reducing material consumption in pursuit of ultimately offering superior solutions to customers.

Moving on to machine tool frames, these elements provide fundamental structural support for every machine tool. Their production requires expensive engineering, testing and high precision manufacturing. As conventionally casted or welded structures, machine tool frames are limited in terms of reconfiguration and cannot be altered after manufacturing. This restricts the reuse to configurations which were initially incorporated into the planning on the part of the engineers.

As the public is becoming increasingly aware of the issue of sustainability, sustainable product manufacturing has become a selling point of its own right. This is true for products manufactured with machine tools, but also for machine tools as products themselves. Previous manufacturing paradigms aimed at producing homogenous products at highest qualities and lowest costs. Nowadays, consumers are demanding the production of individualised goods. This manufacturing paradigm is called mass customisation, which among other developments, most recently led to research on reconfigurable machine tools (RMT) within reconfigurable manufacturing systems (RMS).

Reconfigurable Machine Tools Reconfigurable Manufacturing was defined by Koren et al. (1999), and pointed towards the need for scalable and adaptable manufacturing equipment. One solution to enable the necessary shifts in manufacturing paradigms is the introduction of RMT, made from different modules. With this concept, the foundations of the modular design of machine tools were studied extensively and compiled by Ito (2008). Pioneer work in the modularity of machines was done by Herrmann and Brankamp (1969), who defined the idea of Building Block Systems (BBS). Since then, many research and industrial activities regarding modularisation and reconfiguration in manufacturing (tool design) have been carried out, some of which have found their way into industrial application.

Mori and Fujishima (2009) have presented designs of reconfigurable CNC machine tools addressing the design concept, machine tool configuration and application examples, respectively. The design of the machine tools allowed for selecting a number of axes by the individual axis modules and reconfiguration of the spindle in horizontal and vertical directions. Wulfsberg et al. (2013) give a summary of the concepts developed in the context of modularity in small machine tools for micro-production. The design measures are associated with those of conventional machine tool components for the development of modular systems. Scalability of the working area, namely the change in size of the working area, was achieved within the research work presented. Abele and Wörn (2009) describe a catalogue of components developed for reconfigurable machine tools in the project METEOR. A new approach was developed with the "Reconfigurable Multi technology Machine tool" (RMM) concept that enables the integration of multiple production functions in a single workspace. Most related to sustainability issues, the German project LOeWe (German acronym for Life cycle Oriented development of machine tools) aimed at designing a modular machine tool capable of serving as the basis for different manufacturing processes by including aspects like use-phases and the corresponding life-cycles (Denkena et al. 2006).

With a higher degree of modularisation, challenges have arisen particularly for modularised machine tool frames due to the decreased stiffness which results. The mechanical module interfaces represent serial compliances, reducing the overall rigidity of the given structure assembled. Of course, rigidity is one of the key factors for high productive manufacturing. At the same time, however, the stability of a machine tool mainly depends on a sufficient level of dynamic stiffness of the frame. A common approach for improving the dynamic behaviour is found in the inclusion of actuators and control loops within machine tool structures for the purpose of enhancing damping or for decreasing the dynamic compliance. A building block system for modular machine tool frames therefore requires individualised sensors and actuators.

In addition to sensor technology (see section Microsystems technology), actuators are of great interest when designing for sustainability. As the paradigm of sustainable product design has emerged over the last decades, the design of actuator systems needs to take multifaceted aspects of sustainability into consideration accordingly. This includes avoidance or at least reduction of energy consumption, the substitution of hazardous materials with environment-friendly ones, and low-cost solutions for the production of actuators, leading to solutions for Green Engineering and Manufacturing (Dornfeld 2012).

Various research work on this topic exists, e.g. analysing the energy efficiency of hydraulic, pneumatic and piezoelectric actuators and improving on those actuators (Eriksson 2007; Harris et al. 2014). Existing approaches for achieving higher efficiency vary depending on the actuating principles involved. The most common approach for improving sustainability has turned out to be downsizing, featuring a combination and reconstruction of systems. A combination of different actuation principles, on the other hand, improves the energy efficiency of actuators and combines the advantages of both principles. In this context, for example, hybrid drives are being designed which provide lower energy consumption than regular linear motor direct drives and combine their higher speed and accuracy with the higher damping of screw drives (Okwudire and Rodgers 2013). Chen et al. (2014) present the design of a novel three-degrees-of-freedom linear magnetic actuator which increases the damping and static stiffness of flexible structures during machining. The actuator uses electromagnetic materials which allow larger load capacity and almost no hysteresis compared to piezo and magnetostrictive materials such as Law et al. (2015) report on a novel electro-hydraulic actuator that attenuates and isolates ground motion to keel dynamic excitations transmitted to machine tools below permissible levels. The analysis of optimal placement of actuators can also lead to increased efficiency and thus has to be taken into consideration (Okwudire and Lee 2013).

Applying this principle into the context of RMT, sustainability benefits are anticipated by designing building block systems using tiered technological architectures. Passive lightweight modules can be used to provide structural integrity. Meanwhile, adaptable and reusable sensor technology can increase the smartness of the building block system and improve the overall machine tool frame performance in combination with actuating modules and closed loop controls algorithms.

2.3 Adaptronics in Machine Tools

One of the major limiting factors for the machining quality at high cutting speeds is the static and dynamic behaviour of a machine tool (Ast et al. 2007). The challenge of achieving high static and dynamic stiffness and implementing lightweight design requires adaptronic solutions, which allow for the direct influence of the structural properties of mechanical structures. Adaptronic systems can be integrated into machine tools for different purposes, e.g. active error compensation, active vibration control and active chatter avoidance. The integration of adaptronic systems into machine tools stands as a (key) enabler for achieving higher machining performance as well as for reducing resource consumption, emissions and costs in manufacturing.

A key example of such adaptronic systems is found in dual-stage feed drives. **Dual-stage feed drives** designed for the purpose of allowing high precision positioning over a large workspace on conventional machine tools, fast tool servos (FTS) (e. g. piezoelectric actuator driven flexures) are connected in a series with a machine tool drive in a so-called dual-stage feed drive (DSFD) setup. Woronko et al. (2003) implemented a piezo-based FTS for precision turning on conventional CNC lathes. The results show that the tool positioning accuracy as well as the surface quality could be increased. Elfizy et al. (2005) investigated DSFD for milling processes in pursuit of enabling high precision positioning over a large workspace. In that process, a two-axis flexure mechanism featuring piezoelectric actuators is connected in a series with the machine tool drive stage. The tracking error for sinusoidal profile milling was reduced by approximately 80 % compared to a single stage feed drive. In addition, Drossel et al. (2014) show versatile applications of adaptronic systems in machining processes. FTS systems are currently applied in the form of honing processes for the purpose of increasing the positioning accuracy and for reducing vibration of the tool. The achievable shape accuracy could be therein improved to ± 3 µm and the surface roughness decreased to a reduced peak height of $R_{pk} = 1.7$ µm (Drossel et al. 2014).

Active vibration control
Although active vibration control is not a new concept, a recent development in the field of chatter avoidance for machine tools by means of active damping was discussed by Brecher et. al. (2013). Hömberg et al. (2013) investigated the influences on chatter and solutions for chatter avoidance to improve the efficiency of

production of high quality parts at higher removal rate. Ast et al. (2007) integrated an adaptronic rod in a lightweight structure of a lambda kinematics machine tool in order to overcome vibrations at the tool centre point (TCP), which were identified as a limiting performance factor. Moreover, the active component is designed in a modular manner in such a way that it is transferable to comparable machine tools.

Structure integrated adaptronic components were introduced by Brecher and Manoharan (2009). These devices can compensate deformations of slider structures. Quasi-static and dynamic compensation can be designed for translational and/or rotational axes (Abele et al. 2008; Aggogeri et al. 2013). The modularity is addressed by designing a single unit, which can be used as an active workpiece holder or as a device mounted on a spindle for vibration control (Aggogeri et al. 2013). Chen et al. (2014) presented a smart way of orienting electromagnetic actuators by obtaining two translational and one rotational degree of freedom for an active workpiece holder. Real time compensation of geometric deviations is provided by employing rigid body simulations implemented in the form of an observer in CNC-control (Denkena et al. 2014).

Control strategies With developments in control theory, sophisticated and 'easy to implement' control strategies have evolved over time. To that end, Tiwari et al. (2015) presented an investigation of the application of artificial intelligence techniques, such as fuzzy logic, neurofuzzy, genetic algorithm, genetic programming and data mining in the mechanical engineering domain. An Artificial Neural Network (ANN)-based system of identification and control of dynamic systems was proposed in the late 1980s and early 1990s (Narendra and Parthasarathy 1990). Thereafter, many applications based on neural network control were developed. For instance, a control based on two neural networks with a radial basis function was proposed by Liu and Fuji (2014) for precise positioning of a system with piezoelectric actuators.

Upgrade on demand

The requirements on machine tools depend on and change with the manufacturing task at hand, which is a challenge in a low batch size production environment, see Fig. 1. Machine tools which do not meet the required properties are considered outdated and need a technical overhaul in order to produce productively.

In the machine tool sector, retrofitting is a common principle in the pursuit of reviving outdated machine tools. Retrofitting is primarily understood as steady modification of an existing machine tool and comprises activities such as turning-off or replacing components of machine tools to save energy, or the exchanging out of key wear parts (Gontarz et al. 2012). These activities target a machine tool, its auxiliary systems or the machining process itself.

Among the concept add-ons presented here are optional equipment for upgrading specific functions of machine tools in a flexible manner. The conceptual distinction between retrofit and upgrade by means of add-ons is that the application of add-ons is not permanent but flexible to the required specifications. Sharing add-ons for a pool of machine tools thus enables a production environment to be more resource efficient.

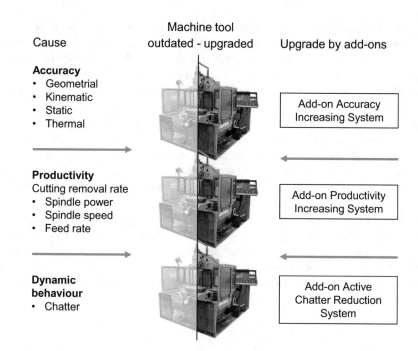

Fig. 1 Concept of machine tool upgrade by add-ons

Easy-to-install capabilities are a necessary feature of add-ons in order to apply the systems in a flexible manner. In addition to the mechanical fixing (e.g. by releasable fastener) on a machine tool, the add-ons applied should be independent of the machine tool control thus allowing the upgrading of a wide range of machines.

2.4 Dry Machining

Whenever cooling liquid (CL) is used in the production process, tools and work-pieces are cooled directly and the friction between the two of them is reduced. This leads to improved tool life and a better workpiece surface quality at the same time. Another advantage is that all chips are washed out of the working area (Klocke and Eisenblatter 1997). If CL is employed in the production process, the following disadvantages have to however be accepted—usually, the chips are contaminated with CL and must be cleaned in downstream installations such as centrifuges and briquetting machines. Moreover, occupational-safety measures have to be applied in order to limit CL's harmful effects to workers' health (Klocke and Eisenblatter 1997; Byrne et al. 2003; Heisel and Lutz 1993a, b).

Dry processing is a production technique characterised by low-power consumption and lean manufacturing chains, as there is no need for the production, monitoring and disposal of cooling lubricants (Ward et al. 2016). This both saves on downstream cleaning and eliminates potential working time lost due to sickness caused by contact with cooling lubricants. However, no cooling of the cutting edge takes place, and the positive effects of lubrication in an interrupted cutting process are likewise lost. In a worst-case scenario, a switchover to dry processing will involve downward revision of the wet processing cutting parameters, which means that the savings on cooling lubricants and the benefits of lean manufacturing processes are offset by a loss in productivity. Alternatively, cooling can be affected with compressed air, solid carbon dioxide or liquid nitrogen (Uhlmann et al. 2012). These alternative cooling agents volatilise in the machining zone, yet their application comes with such high costs that only in exceptional cases do they stand as economically reasonable options (Uhlmann et al. 2012, 2016). Another low-cost tool cooling solution is found in closed internal cooling systems which use a heat sink for dispersing the tool's machining heat in a cooling medium. The accumulated heat in the tool is directed away, and thus is separated from the rise in temperature during actual cutting time.

3   Sustainable Solutions

This chapter describes the development of three sustainable production technology concepts as shown in Fig. 2, schematically. The concepts and solutions for sustainable manufacturing technology presented in Sect. 2 require a step-by-step change within this discipline. In this process, the first step to be realised lies in the integration of innovative machining processes into existing machine tools. In the second step, the modularization of machine tool frames is to be realised, taking into account smart microsystem technology and innovative machining processes. In the final step, the upgrading of the machine tools can be undertaken as necessary.

Internally-cooled tools Turning tools with closed internal cooling systems must meet a different set of requirements compared to conventional tools. In particular, they must ensure mechanical stability under the temperature range of the deployed cooling medium. An internally-cooled turning tool and cooling periphery was, furthermore, developed within the framework of the work described here.

LEG^{2}O smart building block system As Koren et al. (1999) stated, RMT are anticipated to contribute significantly to the manufacturing of mass-customised products. Key attributes related to the sustainability of RMT result from delivering the machining purpose when and as needed and from avoiding downtime or underutilization. Although modularization was achieved on the hardware side of machine tools, among various other components in the past, the modularisation of the machine tool frame remains an open question of great relevance. A Smart Building Block System (BBS) for modular machine tool frames is envisioned to overcome technological limitations and to provide a sustainable alternative to the

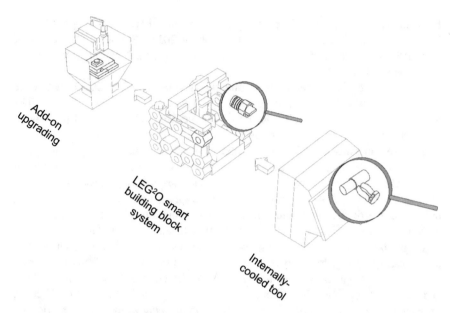

Fig. 2 Sustainable machine tool concepts—development from the present to the future

design of conventional machine tools. This is achieved by using a tiered techno-logical approach combining microsystem technology (MST) and mechatronic technology (MT).

Add-ons for outdated machine tools Compared to current machine tools, outdated machine tools suffer from high positioning deviations and fail to add sufficient value to current production systems (Uhlmann and Kianinejad 2013). Under this concept, this problem is tackled by an upgrade using an Add-on Accuracy Increasing System (AAIS). The add-on systems are used without exchanging essential system components. By upgrading or enhancing the func-tionality of outdated machine tools rather than replacing them with new machines, valuable resources can be saved (Allwood et al. 2011). The add-on solution developed, furthermore helps to keep outdated machine tools competitive in the contemporary production chain by increasing not only positioning accuracy but also by allowing for active vibration control to reduce chatter (Kianinejad et al. 2016).

3.1 Smart Building Block Systems

The technological basis of the LEG^{2}O BBS (German acronym for lightweight and accuracy optimised) consists of passive, active and MST enhanced smart building blocks, called modules, with a weight limitation of 30 kg per unit. In accordance with worker safety regulations for manual lifting, each component of the LEG^{2}O

BBS allows for manual handling during assembly, maintenance and upgrading with low-level infrastructure (Steinberg and Windberg 2011). Studies on use-cases of these BBS led to the conclusion that for LEG^{2}O BBS, most cutting machine tool scenarios are sound (Peukert et al. 2013).

What's more, first order impacts directly associated with LEG^{2}O BBS were addressed in a tiered life cycle sustainability assessment (Peukert et al. 2015b). The research revealed clear benefits during the whole life cycle compared to conventional static machine tool frames due to reuse and adaptation of machine tools.

Passive modules provide fundamental mechanical properties, e.g. rigidity, to support the core structure of modular machine tool frames. Within the scope of sustainable design, resource efficiency was set as a target during the design phase of the passive module. A bionic-inspired fractal design approach was chosen, resulting in high geometric flexibility with stiff and scalable structures based on two modules with a hexagonal prism shape (Peukert et al. 2015a). These modules were topologically optimised to provide a lightweight and resource-efficient foundation with the necessary rigidity to compete with conventional machine tool frames. Figure 3 shows the development history of the different design stages of passive modules and shows the topologically optimised hexagons with a weight of 6.2 and 12.3 kg, respectively. Yet, on top of advantages in damping behaviour, decreased rigidity and dynamic stiffness are caused by the high number of joints (Uhlmann and Peukert 2015), leading to the need for active modules and compensation.

Smart modules The physical instantiation of MST by means of wireless sensor nodes, provides smart functionalities (e.g. orientation sensing) in dedicated passive modules within the frame, or so-called smart modules. Data on module identification as well as parameters related to the physical state of the machine tool frame are shared between smart modules and one centralized receiving unit using wireless communication. The demand for fully autonomous concepts and long-term usage called for a revised catalogue of requirements apart from conventional industrial sensors. As the overall system size of highly miniaturised sensors is dominated by the autonomous energy supply, measures for reducing the average power consumption were identified as crucial aspects of system design (Lambebo 2014). With highly efficient programming routines, e.g. reduced instruction sets and the relocation of dedicated tasks into the base stations, operating times were even further prolonged without the need for battery replacement. However, only through the application of advanced packaging technologies can optimised form factors be

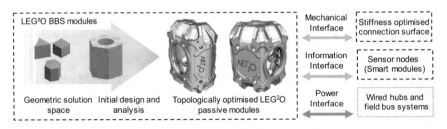

Fig. 3 Topologically optimised passive modules and interfaces

achieved that allow for the least interference at maximum functional density and long lifetime at the workstation. A combined approach including Flip Chip assembly, surface mount technology and a sequence of embedding processes on panel level was furthermore used to realise a new generation of WSNs. Physical devices demonstrating the fusion of MST hardware with MT (e.g. in a screw) are shown in Fig. 4 with selected functional groups indicated.

Active modules are used to compensate for displacements. The control loop of the actuators uses on-line data, transmitted wirelessly by smart modules. Apart from the machine tool dynamic, thermal loads constitute a main cause of inaccuracies with machine tool frames. Hence, the active modules have a control platform manipulated by three separate compliant mechanisms. The compliant mechanisms are driven by the thermal deformation of aluminium bars controlled by thermo-electric modules. The slow nature of thermal deviations allows the usage of solid state relays to power numerous active modules from a single power supply. Hence, additional sustainability benefits can be achieved in comparison to traditional piezo-driven approaches, which require one amplifier per channel. At a simplified level, the construction and usability of this approach, as well as the concomitant sustainability benefits and control of the active module, were analysed (Uhlmann and Peukert 2015). The topology of the mechanism facilitates a self-adapting passive compensation movement at the output platform by change of ambient temperature due to an inherent thermal compensation of approximately $x_i \approx 2$ μm/K. The actuator is designed to provide a compensating range of $\Delta x = 100$ μm maintaining micrometer accuracy. Figure 5 shows the actuator and experimental results of the closed loop control in a prototypical test structure.

A combined hard- and software infrastructure synchronises all relevant data for the analysis of thermal distortion of the frame as shown in Fig. 6. Additional data sets, e.g. module identification and orientation, are distributed to external devices for visualisation on a tablet PC. A methodology was developed to support the spatial setup of WSNs within the LEG^{2}O BBS (Uhlmann et al. 2014). This method supports the optimisation of the relationship between the mechanical structure,

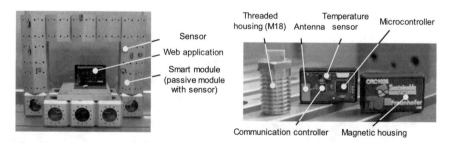

Fig. 4 Wireless sensor nodes for parameter monitoring of passive LEG^{2}O modules—the final technology demonstrator of the sensor fits into a hollow M18 screw

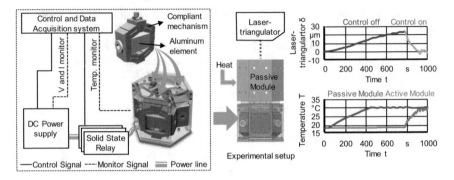

Fig. 5 Active module for compensation of thermal deviations

loading scenario and the number of sensor nodes, and thus helps minimize the utilization of MST.

3.2 Add-Ons for Machine Tool Upgrade

Outdated machine tools suffer from high positioning deviations compared to current machine tools and fail to add sufficient value to present production systems (Uhlmann and Kianinejad 2013). Within the scope of the work described here, this technical obsolescence is addressed by an upgrade using an Add-on Accuracy Increasing System (AAIS). The add-on systems are used without exchanging essential system components. By upgrading or enhancing the functionality of outdated machine tools rather than replacing them with new machines, valuable resources can be saved (Allwood et al. 2011). The add-on solution developed helps to keep outdated machine tools competitive in the contemporary production chain by increasing not only positioning accuracy, but also by allowing for active vibration control to reduce chatter (Kianinejad et al. 2016).

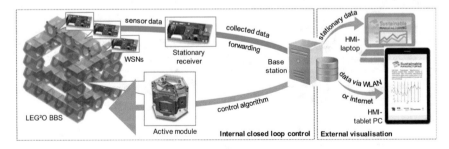

Fig. 6 Concept of the smart modules equipped with MST (WSNs) and IT infrastructure (Base station, HMI)

The energy efficiency for different machining operations of an outdated milling machine and a modern one was compared in (Kianinejad et al. 2015). Though the energy consumption of the outdated machine exceeds that of the newer one, the upgrade of an outdated machine tool presented saves raw materials, energy for material extraction, along with manufacturing energy by not replacing the outdated milling machine with a new one.

Add-on Accuracy Increasing System (AAIS) A high accuracy error compensation table has been integrated and tested on a representative milling machine tool, shown as Fig. 7. The FP4NC milling machine, FRIEDRICH DECKEL AG, München, Germany is run by a GRUNDIG Dialog 4 control, so that compensation by control unit is not possible. Sensors of the add-on system can measure a significant portion of static, dynamic, and kinematic inaccuracies. These measurements are used, together with a feedback control mechanism to correct the errors by means of piezoelectric actuators. The add-on error compensation table is run by separate control hardware (dSPACE 1103) and does not share the control unit of the machine tool, making the solution developed both modular and independent of the type of machine. In order to use the error compensation table depending on the respective manufacturing tolerances and to allow sharing with different machine tools, the main challenge with the integration of an error compensation table is to provide easy-to-install capabilities. These capabilities are achieved by independent control, as only one interface is required to the machine tool for sensor readouts, along with a screw connection of the table, see also (Kianinejad et al. 2016).

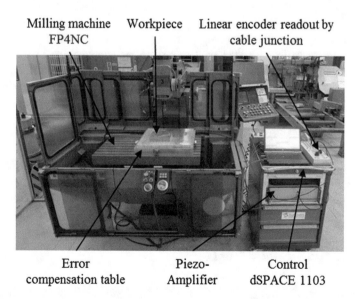

Milling machine   Workpiece   Linear encoder readout by
FP4NC   cable junction

Error   Piezo-   Control
compensation table   Amplifier   dSPACE 1103

Fig. 7 Test setup of a upgraded milling machine FP4NC, FRIEDRICH DECKEL AG, München, Germany

Fig. 8 Error compensation
table (clamping plate
removed)

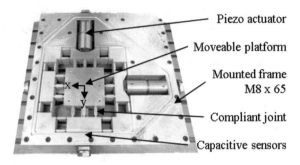

Piezo actuator

Moveable platform

Mounted frame
M8 x 65

Compliant joint

Capacitive sensors

The error compensation table corrects the relative position between workpiece and tool in real time and provides compensation in two perpendicular axes in the horizontal plane (see Fig. 8). A capacitive sensor mounted in the frame is used in each direction to detect the motion of the platform with respect to the frame. Along with measuring the quasi-static position, these sensors also measure the dynamic movement of the platform. Together with the piezo actuators and control designed, active damping is also provided by AAIS.

The piezo actuators are pre-stressed by a housing in order to protect them against forces. The nominal stroke of the piezo actuators $s_A = 125$ μm is reduced by the applied pre-stress and voltage of amplifiers, so that AAIS can provide a compensation of $s_{AAIS} = 55$ μm in each of the axes.

To overcome the problem of stick-slip and backlash, the error compensation table is designed to be monolithic, featuring compliant joints which provide high stiffness up to $k_z = 100$ N/μm in the vertical direction (Kianinejad et al. 2016). In order to provide high strain by low stress in the compliant joints, aluminum alloy (AW7075) was chosen due to its high ratio of strength to elastic modulus.

A look-up table is generated containing the repeatable error by initial measurements on the linear positioning of the machine in the x and y axes. By feeding the look-up table with the reference position (x, y) of the linear encoders of FP4NC, FRIEDRICH DECKEL AG, München, Germany a reference signal is generated. This reference signal is then fed to the control of the xy-table for tracking.

Figure 9 shows the static positioning test performed by a laser interferometer on the x axis of FP4NC with and without the AAIS and also compares the data to the positioning accuracy of a modern machine tool, a DMU50, DMG MORI AG, Bielefeld, Deutschland produced in 2008. It can be seen that AAIS improves the positioning accuracy significantly.

3.3 *Internally-Cooled Tools*

The **tool design** of turning tools was reconsidered when taking the requirements and conditions of an integrated closed-loop cooling system for turning tools into

Fig. 9 Positioning deviations of the x-axis

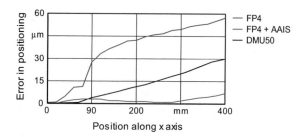

consideration, e.g. cooling medium temperatures from −210 to +40 °C. As a result, topology optimised tool geometries were simulated and investigated. The resulting tool design concepts were assessed by taking into account the weight, stiffness, and the level of integration (Uhlmann et al. 2014). The position of the cutter should likewise avoid shifting during the operation of the cooling system, as this otherwise results in marked variations in the geometry of the finished components. To manage this, the cooling medium channels are integrated into the turning tool holder body and are decoupled from its support structure. Manufacturing of flow-optimised cooling channels is difficult, thus selective laser melting is used for fabrication of the tool holder, see Fig. 10.

Comparison of different cooling strategies A comprehensive analysis of three different cooling methods compared to a dry machining process was carried out with a variation of different cutting speeds v_c:

- Dry machining with an internally-cooled tool with a water-ethanol mixture as process coolant
- Dry machining with an internally-cooled tool with liquid nitrogen as process coolant
- Flood cooling

The energy demand of these cooling methods compared to a dry machining process is given in Fig. 11. The use of liquid nitrogen as a process coolant improves the tool lifetime by at least 50 %. However, due to the temperature influence of the coolant, the TCP shows a displacement of up to dT = 0.2 mm. The total energy

Fig. 10 Internally-cooled turning tool and cooling periphery

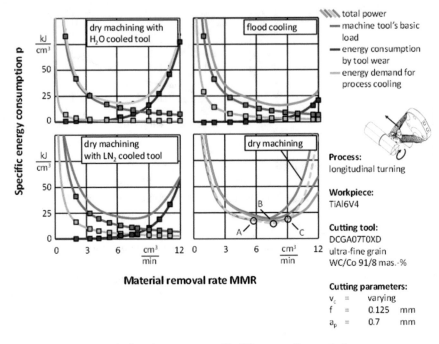

Fig. 11 Energy demand of cutting processes with different cooling methods

demand is comparable to flood cooling. Using the water-ethanol mixture, the tool lifetime can be improved by at least 40 % while the TCP shows no thermal displacement. In addition, the process yields the highest energy efficiency (see point B in Fig. 11).

In summary, the need for process and tool cooling depends on the chosen process parameters. The highest energy efficiency for part finishing can be achieved with dry machining (point A), for semi-finishing with indirect cooled tools (point B), and for very high material removal rates with flood cooling (point C).

4 Conclusion

To conclude, this chapter shows the ongoing research in pursuit of a sustainable impact on the worldwide production industry by means of the development of innovative solutions for manufacturing environments. A special focus is placed on concepts for reconfigurable machine tools: where fluctuating production environments cannot be tackled by conventional static constructions, a modular building block system is designed that enables the production of homogenous products at high quality standards, lowest costs and featuring the option for partial replacement, repair, exchange, or upgrade with regular service intervals, therein avoiding downtime or underutilization. The BBS concept encompassed various perspectives for overcoming technological limitations. The technological basis of the proposed LEG^{2}O BBS consists of a scalable structure, where connections, interfaces and microsystem technology constitute elements for connecting and enabling the application of two basic module geometries of a hexagonal prism shape, involving active modules that allow for the compensation of thermal deformations. In this process, the actuator of different actuation principles as well as optimal placement within the machine tool frame serve together to improve energy efficiency and combines the advantages of the principles. In addition, passive modules provide stiffness and serve as the structural base.

Smart modules are enhanced by microsystem technology, namely wireless sensor nodes. Battery and energy harvesting technology in combination with highly efficient programming routines and customised hardware architecture allow for autonomous and flexible hard- and software infrastructure to synchronise all relevant data for the analysis of thermal distortion. Adaptive components then increase the accuracy of machine tools while remaining up-to-date.

The increase of the accuracy of machine tool frames, the usage of mechatronics regarding the accuracy of axes, as well as the "sustainable engineering" of solutions and recycling of components and equipment constitute the main parts of the research work at hand. Detail questions for the future concern communication reliability, data losses, environmental trade-offs between benefits and impact, as well as evaluation approaches for 25+ years of use time. A link between the microsystems' impact on component level and modularity is so far still missing. Moreover, long-term compatibility between the different lifecycles and concepts of machine tool components and electronics needs to be investigated. Though modularisation was achieved on the hardware side of machine tools among various components, the net effect on sustainability of the modularisation of the machine tool frame remains an open question.

By upgrading outdated machine tools with add-on components, the accuracy of modern machine tools is achieved. Sensors of the add-on system can measure a significant portion of static, dynamic, and kinematic inaccuracies. These measurements are applied together with a feedback control to correct the errors through the use of piezoelectric actuators.

By upgrading or enhancing the functionality of older machine tools, valuable resources and energy consumption can be saved. This aim is realised without exchanging essential system components, but by using high precision compensation add-on systems. Sensors of the add-on system are able to measure a portion of static, dynamic and kinematic inaccuracies. The add-on compensation is realised by a separate control unit, making the development solution modularised and easily incorporated into different types of machine tools.

References

Abele, E., and A. Wörn. 2009. Reconfigurable machine tools and equipment. In *Changeable and reconfigurable manufacturing systems*, 111–125. Springer.

Abele, E., H. Hanselka, F. Haase, D. Schlote, and A. Schiffler. 2008. Development and design of an active work piece holder driven by piezo actuators. *Production Engineering* 2(4): 437–442. doi:10.1007/s11740-008-0123-3.

Aggogeri, F., F. Al-Bender, B. Brunner, M. Elsaid, M. Mazzola, A. Merlo, D. Ricciardi, M. de la O Rodriguez, and E. Salvi. 2013. Design of piezo-based AVC system for machine tool applications. *Mechanical Systems and Signal Processing* 36(1): 53–65. doi:10.1016/j.ymssp.2011.06.012.

Allwood, J.M., M.F. Ashby, T.G. Gutowski, and E. Worrell. 2011. Material efficiency: A white paper. *Resources, Conservation and Recycling* 55(3): 362–381. doi:10.1016/j.resconrec.2010.11.002.

Ast, A., S. Braun, P. Eberhard, and U. Heisel. 2007. Adaptronic vibration damping for machine tools. *CIRP Annals—Manufacturing Technology* 56(1): 379–382. doi:10.1016/j.cirp.2007.05.088.

Beeby, S. 2006 Energy harvesting vibration sources for wireless sensor networks with special focus on vibrations.

Brankamp, K., and J. Herrmann. 1969. Baukastensystematik—Grundlagen und Anwendung in Technik und Organisation. *Ind.-Anz* 91: 693–697.

Brecher, C., and D. Manoharan. 2009. Aktive Dämpfung für Portalmaschinen: Entwicklung adaptronischer Kompensationsmodule für Schieberstrukturen. *wtWerkstattstechnik online* 99 (5): 288–293.

Brecher, C., S. Bäumler, and B. Brockmann. 2013. Avoiding chatter by means of active damping system for machine tools. *Journal of Machine Engineering* 13(3): 117–128.

Byrne, G., D. Dornfeld, and B. Denkena. 2003. Advanced cutting technology. *Annals of the CIRP-Manufacturing Technology* 52(2): 483–507.

Chen, F., X. Lu, and Y. Altintas. 2014. A novel magnetic actuator design for active damping of machining tools. *International Journal of Machine Tools and Manufacture* 85: 58–69. doi:10.1016/j.ijmachtools.2014.05.004.

Denkena, B., et al. 2006. Life-cycle oriented development of machine tools. In *13th cooperative institutional research program international conference on life cycle engineering*, 693–698.

Denkena, B., L. Overmeyer, K.M. Litwinski, and R. Peters. 2014. Compensation of geometrical deviations via model based-observers. *The International Journal of Advanced Manufacturing Technology* 73(5–8): 989–998. doi:10.1007/s00170-014-5885-5.

Dornfeld, D.A. 2012. *Green manufacturing: fundamentals and applications*. Springer Science & Business Media.

Drossel, W.-G., A. Bucht, C. Hochmuth, A. Schubert, A. Stoll, J. Schneider, and R. Schneider. 2014. High performance of machining processes by applying adaptronic systems. *Procedia CIRP* 14: 500–505.

Elfizy, A.T., G.M. Bone, and M.A. Elbestawi. 2005. Design and control of a dual-stage feed drive. *International Journal of Machine Tools and Manufacture* 45(2): 153–165. doi:10.1016/j.ijmachtools.2004.07.008.

Elvin, Niell, Erturk, Alper (Eds.). 2013. Advances in energy harvesting methods. doi:10.1007/978-1-4614-5705-3.

Eriksson, B. 2007. Control strategy for energy efficient fluid power actuators. Lic Thesis, LiU-Tryck, Linköping.

Gontarz, A., F. Hänni, L. Weiss, and K. Wegener. 2012. Machine tool optimization strategies: Evaluation of actual machine tool usage and modes. In *Sustainable manufacturing*, ed. Günther Seliger, 131–36. Berlin: Springer.

Gungor, V.C., and G.P. Hancke. 2009. Industrial wireless sensor networks: Challenges, design principles, and technical approaches. *IEEE Transactions on Industrial Electronics* 56(10): 4258–4265.

Harris, P., S. Nolan, and G.E. O'Donnell. 2014. Energy optimisation of pneumatic actuator systems in manufacturing. *Journal of Cleaner Production* 72: 35–45.

Heisel, U., and M. Lutz. 1993a. Probleme der umwelt- und humanverträglichen Fertigung am Beispiel der Kühlschmierstoffe – Erster Teil, DIMA, 93/8.9:81–83.

Heisel, U., and M. Lutz. 1993b. Probleme der umwelt- und humanverträglichen Fertigung am Beispiel der Kühlschmierstoffe – Zweiter Teil. *DIMA* 93(10): 35–40.

Hömberg, D., E. Uhlmann, O. Rott, and P. Rasper. 2013. Development of a stability prediction tool for the identification of stable milling processes. In *Process machine interactions*, ed. B. Denkena, and F. Hollmann, 203–224. Springer.

Hu, P. 2015. A system architecture for software-defined industrial Internet of Things. In *IEEE international conference on ubiquitous wireless broadband (ICUWB)*. Paris: IEA Publications.

International Energy Agency. 2007. *Tracking industrial energy efficiency and CO_2 emissions*. Paris: IEA Publications.

Ito, Y. 2008. *Modular design for machine tools*. McGraw Hill Professional.

Kianinejad, K., E. Uhlmann, and B. Peukert. 2015. Investigation into energy efficiency of outdated cutting machine tools and identification of improvement potentials to promote sustainability. Procedia CIRP 26: 533–538.

Kianinejad, K., S. Thom, S. Kushwaha, and E. Uhlmann. 2016. Add-on error compensation unit as sustainable solution for outdated milling machines. *Procedia CIRP* 40: 174–178.

Klocke, F., and G. Eisenblatter. 1997. Dry cutting. *Annals of the CIRP-Manufacturing Technology* 46(2): 519–526.

Koren, Y., et al. 1999. Reconfigurable manufacturing systems. *CIRP Annals-Manufacturing Technology* 48(2): 527–540.

Lambebo, A., and S. Haghani. 2014. A wireless sensor network for environmental monitoring of greenhouse Gases. In *ASEE 2014 zone I conference*.

Law, M., et al. 2015. Active vibration isolation of machine tools using an electro-hydraulic actuator. *CIRP Journal of Manufacturing Science and Technology* 10: 36–48.

Liu, D., and F. Fuji. 2014. An adaptive internal model control system of a piezo-ceramic actuator with two RBF neural networks. In *2014 IEEE international conference on mechatronics and automation (ICMA)*, 210–215. doi:10.1109/ICMA.2014.6885697.

Mori, M., and M. Fujishima. 2009. Reconfigurable machine tools for a flexible manufacturing system. In *Changeable and reconfigurable manufacturing systems*, 101–109. Springer.

Narendra, K.S., and K. Parthasarathy. 1990. Identification and control of dynamical systems using neural networks. *IEEE Transactions on Neural Networks/A Publication of the IEEE Neural Networks Council* 1(1): 4–27. doi:10.1109/72.80202.

Okwudire, C., and J. Lee. 2013. Minimization of the residual vibrations of ultra-precision manufacturing machines via optimal placement of vibration isolators. *Precision Engineering* 37(2): 425–432.

Okwudire, C., and J. Rodgers. 2013. Design and control of a novel hybrid feed drive for high performance and energy efficient machining. *CIRP Annals-Manufacturing Technology* 62(1): 391–394.

Ovsthus, K., and Lars M. Kristensen. 2014. An industrial perspective on wireless sensor networks —a survey of requirements, protocols, and challenges. *IEEE Communications Surveys & Tutorials* 16(3): 1391–1412.

Peukert, B., J. Mewis, M. Saoji, E. Uhlmann, S. Benecke, R. Thomasius, N. F. Nissen, and K.-D. Lang. 2013. Microsystem enhanced machine tool structures to support sustainable production in value creation networks.

Peukert, B., M. Saoji, and E. Uhlmann. 2015a. An evaluation of building sets designed for modular machine tool structures to support sustainable manufacturing. *Procedia CIRP* 26: 612–617.

Peukert, B., S. Benecke, J. Clavell, S. Neugebauer, N. F. Nissen, E. Uhlmann, K.-D. Lang, and M. Finkbeiner. 2015b. Addressing sustainability and flexibility in manufacturing via smart modular machine tool frames to support sustainable value creation. *Procedia CIRP* 29: 514–519.

Rault, T., A. Bouabdallah, and Y. Challal. 2014. Energy efficiency in wireless sensor networks: A top-down survey. *Computer Networks* 67: 104–122.

Schischke, K. 2009. *Innovations- und Technikanalyse autonomer verteilter Mikrosysteme: Schlussbericht.* Berlin, Hannover: Technische Informationsbibliothek u. Universitätsbibliothek.

Sheltami, T., A. Al-Roubaiey, A. Mahmoud and E. Shakshuki. 2015. A publish/subscribe middleware cost in wireless sensor networks: A review and case study. In *IEEE 28th Canadian conference on electrical and computer engineering (CCECE).*

Steinberg, U., and H.-J. Windberg. 2011. *Heben und Tragen ohne Schaden.* Dortmund: Baua.

Tiwari, A., R. Jaideva, and S. K. Pradhan. 2015. An investigation into use of different soft artificial intelligence techniques in mechanical engineering domain. In *Proceedings of fourth international conference on soft computing for problem solving*, ed. K.N. Das, K. Deep, M. Pant, J.C. Bansal and A. Nagar, 521–541. Springer.

Uhlmann, E., and K. Kianinejad. 2013. Investigation of the upgrading potentials of out-of-date cutting machine tools to promote sustainable and global value creation. In *Proceeding of the 11th global conference on sustainable manufacturing*, ed. G. Seliger, 574–579. Berlin: University Publisher of TU Berlin.

Uhlmann, E., and B. Peukert. 2015. Erhöhen der Dämpfung hohler Leichtbaustrukturen: Ein passiver und nachhaltiger Ansatz zur Verbesserung der Strukturdämpfung. *Wt Werkstattstechnik online* 105(7/8): 446–450.

Uhlmann, E., P. Fürstmann, M. Roeder, S. Richarz, and F. Sammler. 2012. Tool wear behaviour of internally cooled tools at different cooling liquid temperatures. In *Proceedings of the 10th global conference on sustainable manufacturing*, ed. G. Seliger, and E. Kılıç, 21–27. Berlin: University Publisher of TU Berlin.

Uhlmann, E., M. Saoji, S. Kushwaha, and B. Peukert. 2014. Optimierte Sensorplatzierung in Mess-Netzwerken. *Werkstattstechnik online* 5: 266–271.

Uhlmann, E., M. Saoji, and B. Peukert. 2016. Utilization of thermal energy to compensate quasi-static deformations in modular machine tool frames. *Procedia CIRP* 40: 1–6.

Verein Deutscher Werkzeugmaschinenfabriken e.V. 2015. Marktbericht 2014: Die deutsche Werkzeugmaschinenindustrie und ihre Stellung im Weltmarkt [Online]. http://www.vdw.de/. Accessed 11 April 2016.

Wagner, E., S. Benecke, J. Winzer, N.F. Nissen, and K.-D. Lang. 2016. Evaluation of indicators supporting the sustainable design of electronic systems. *Procedia CIRP* 40: 469–474.

Ward, H., M. Burger, Y. Chang, P. Fürstmann, S. Neugebauer, A. Radebach, G. Sproesser, A. Pittner, M. Rethmeier, E. Uhlmann, J. Steckel. 2016. Assessing carbon dioxide emission reduction potentials of improved manufacturing processes using multiregional input output frameworks. *Journal of Cleaner Production.* doi:10.1016/j.jclepro.2016.02.062.

Woronko, A., J. Huang, and Y. Altintas. 2003. Piezoelectric tool actuator for precision machining on conventional CNC turning centers. *Precision Engineering* 27(4): 335–345. doi:10.1016/S0141-6359(03)00040-0.

Wulfsberg, J. P., A. Verl, K.-H. Wurst, S. Grimske, C. Batke, and T. Heinze. 2013. Modularity in small machine tools. *Production Engineering* 7(5): 483–490.

Zurawski, R. 2009. Wireless sensor network in industrial automation. In *International conference on embedded software and systems, Pittsburgh, 2009.*

Sustainable Technologies for Thick Metal Plate Welding

Gunther Sproesser, Ya-Ju Chang, Andreas Pittner,
Matthias Finkbeiner and Michael Rethmeier

Abstract Welding is the most important joining technology. In the steel construction industry, e.g. production of windmill sections, welding accounts for a
main part of the manufacturing costs and resource consumption. Moreover, social
issues attached to welding involve working in dangerous environments. This aspect
has unfortunately been neglected so far, in light of a predominant focus on economics combined with a lack of suitable assessment methods. In this chapter,
exemplary welding processes are presented that reduce the environmental and
social impacts of thick metal plate welding. Social and environmental Life Cycle
Assessments for a thick metal plate joint are conducted for the purpose of
expressing and analysing the social and environmental impacts of welding.
Furthermore, it is shown that state-of-the-art technologies like Gas Metal Arc
Welding with modified spray arcs and Laser Arc-Hybrid Welding serve to increase
social and environmental performance in contrast to common technologies, and
therefore offer great potential for sustainable manufacturing.

Keywords Life Cycle Assessment (LCA) · Arc Welding · Laser Arc-Hybrid
Welding · Resource efficiency · Social Life Cycle Assessment (SLCA) · Human
health

G. Sproesser (✉) · M. Rethmeier
Institute of Machine Tools and Factory Management, Technische Universität Berlin,
Berlin, Germany
e-mail: gunther.sproessser@googlemail.com

Y.-J. Chang · M. Finkbeiner
Department of Environmental Technology, Technische Universität Berlin, Berlin,
Germany

A. Pittner · M. Rethmeier
Department of Component Safety, Federal Institute for Materials Research and Testing,
Berlin, Germany

R. Stark et al. (eds.), *Sustainable Manufacturing*, Sustainable Production,
Life Cycle Engineering and Management, DOI 10.1007/978-3-319-48514-0_5

1   Introduction

Welding plays a pivotal and irreplaceable role in modern manufacturing. The applications are involved in nearly all industries, for example, construction, automobile, turbine production, etc. Yet welding processes require large amounts of energy and resources which are of course critical from an environmental perspective. Social aspects of welding meanwhile mainly involve health effects associated with welding fumes and welder compensation.

Common welding technologies include Gas Metal Arc Welding (GMAW), Manual Metal Arc Welding (MMAW) and Laser Arc-Hybrid Welding (LAHW), which all differ tremendously in their properties and potential in the realm of sustainable manufacturing.

MMAW with coated electrodes is a popular welding technology on building sites due to the fact that it offers high flexibility and requires no shielding gas supply. Additionally, low costs of equipment and electrodes incentivize the frequent application of MMAW. On the other hand, the productivity attached to MMAW tends to be low due to limited welding speeds, process power capacity limitations, as well as the attendant additional time consumption at play when changing the electrode and removing the slag. Furthermore, MMAW is performed manually, which entails significant health risks for welders.

Meanwhile GMAW is one of the most widely used technologies due to the fact that it is easy to automate and offers a high level of productivity and flexibility. The typical operation mode of GMAW for the purpose of achieving high deposition rates and process speeds, is automatic welding with spray arc transfer. Recently, manufacturers of welding power sources have developed modern arc processes as presented early by Dzelnitzki (2000), and later by Lezzi and Costa (2013). One innovation is a highly concentrated spray arc that enables higher penetration depths and the reduction of flange angles. Consequently, the modern modified spray arcs lead to reduced material consumption which prove to be promising with respect to environmental aspects.

Then there's LAHW, which remains a rather young technology compared to those mentioned above, yet is well on its way as a promising new field of sustainable manufacturing. In comparison with GMAW, LAHW achieves higher welding speeds and hence higher productivity, while the reduced number of passes and lower volume of molten material lead to resource savings, lower distortion and less rework. Yet when it comes to large structures with high geometrical tolerances of several millimetres, gap bridging can be a critical issue ultimately limiting the application of LAHW as it stands.

For manufacturing processes and products, environmental and social issues are often insufficiently considered and respected. The negative effects on the environment and humans however accumulate, many of which are also irreversible. To evaluate the environmental impacts and social influences of a process or product, Life Cycle Assessment (LCA) (ISO 2006a; Schau et al. 2012), and Social Life Cycle Assessment (SLCA) (UNEP 2009) are the current state-of-the-art methodologies. LCA is an ISO standardised method, widely employed for providing an estimate on the potential environmental impacts of products through the whole life cycle (Schau et al. 2012; Klöpffer and Grahl 2009; Guinée et al. 2002). It is the most advanced and tried-and-true methodology in evaluating environmental burden on process or product levels, and also in preventing burden shifting from different life cycle phases.

According to the guidelines for Social Life Cycle Assessment of Products (UNEP 2009), SLCA is defined as a methodology that aims at assessing the potential positive and negative social and socio-economic impacts related to human beings affected by products/services throughout the life cycle, such as health and wage issues of workers, etc. Though SLCA studies have increased in number significantly within the last three years, the method is still considered to be rather in its infancy (Neugebauer et al. 2015).

To date, welding technology developments and comparisons remain predominantly focused on economic indicators. Environmental and social aspects are insufficiently taken into account when evaluating and choosing a process for a given welding task. To that end, MMAW, LAHW and automatic GMAW with a conventional spray arc and a modified spray arc, have been evaluated in view of the environmental and social aspects attached. SLCA and LCA have been applied to compare the corresponding environmental impacts and the potential health risks to welders, particularly caused by welding fumes. Moreover, the wage status of welders in Germany has been investigated with a discussion of the fairness and adequacy given their working and living conditions. The results can help the industry to identify the crucial issues and then offer improvements to the processes and equipment in pursuit of more sustainable alternatives.

2 Methodology

2.1 Environmental Assessment

According to the ISO standard, the methodology is divided into these four phases: goal and scope definition, life cycle inventory analysis, life cycle impact assessment, and interpretation in an iterative process (ISO 2006a, b). First of all, the goals of this LCA study are to highlight the environmental impact contributed by different inputs and outputs of the chosen welding processes, and to compare the differences in environmental impact. The results are expected to provide information for welding process development and selection. The scope of the study is concerned with the welding processes in and of themselves, including the life cycle stages of material acquisition (involving in raw material extraction and processing of the used material in welding processes), the manufacturing phase (carrying out welding processes), and waste management. In line with the defined scope, the system boundary covers the consumption of electricity, materials and gases, and landfill

System boundary

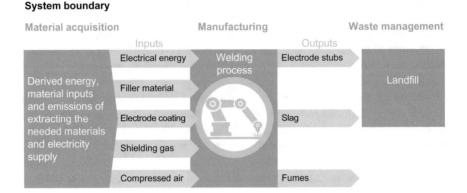

Fig. 1 System boundary and inputs and outputs of welding processes (Sproesser et al. 2015)

waste, but stops short of considering machinery. The functional unit is 1 m weld seam of a 20 mm thick metal plate. The input and output information based on the defined functional unit will be collected and calculated in life cycle inventory analysis stage. In this study, the CML 2002 method is adopted as the life cycle impact assessment method (as the midpoint approach). Meanwhile, GaBi 6.0 (by thinkstep) is used as the software to build and carry out the LCA model.

In the life cycle inventory analysis phase, the inventory data of inputs and outputs of the chosen welding processes are collected according to the system boundary and the functional unit. Figure 1 shows the considered process inputs and outputs, filler material, shielding gas, electrical energy, welding fumes, compressed air (for LAHW), electrode coating (for MMAW), electrode stubs (for MMAW), and slag (for MMAW).

Electricity consumption for the welding processes was determined with values measured and the respective wall-plug efficiency of the equipment. The wall-plug efficiency of arc welding machines (MMAW, GMAW and the arc content of LAHW) was set to 80 % (Sproesser et al. 2016; Hälsig 2014). For LAHW, electricity consumption of the beam source took into account process power, an efficiency of 30 %, and additional contributions of the cooling unit. Electric energy for robot movement was measured at the feed cable for the respective trajectories and added to the electricity demand of the welding source in order to calculate the overall energy utilised for the joining process.

The consumption of filler material was determined by measurement of the wire feed rate and in the case of MMAW, by weighting the electrodes and by collecting the remaining electrode stubs. The chemical compositions of the materials were taken from available product data sheets. For MMAW, only titanium dioxide (45 %) and silicon dioxide (10 %) were considered to represent the main composition of the electrode coating due to missing data in the GaBi data base. The consumption of compressed air for LAHW was estimated by applying Bernoulli's principle to the geometry of the cross-jet unit of the laser head.

Table 1 Fume emission rates of the applied welding processes

	Emission rate
MMAW	4 mg/min (Pohlmann et al. 2013)
GMAW standard	6 mg/s (Rose et al. 2012)
GMAW modified	4 mg/s (Rose et al. 2012)
LAHW	LAHW root pass: 10.4 mg/s (Pohlmann et al. 2013)
	GMAW filler pass: 6 mg/s (Rose et al. 2012)

Fume emissions are calculated according to emission rates of representative processes (power range and transfer mode) from literature (Pohlmann et al. 2013; Rose et al. 2012) and are displayed in Table 1. The chemical composition is assumed to be mainly from iron oxide (Antonini et al. 2006; Jenkins and Eagar 2005).

Considering the robustness, practicality, and the close relationship between welding technologies and metal related industry, the four indicators: global warming potential (GWP), eutrophication potential (EP), acidification potential (AP) and photochemical ozone creation potential (POCP) have been selected for further comparison in life cycle impact assessment stage (World Steel Association 2011; PE International 2014). GWP (100 years, in kg of carbon dioxide equivalent) evaluates the long-term contribution of a substance to climate change. EP (in kg phosphate equivalent) estimates the impact from the macro-nutrients nitrogen and phosphorus in bio-available forms on aquatic and terrestrial ecosystems, affecting undesired biomass production. AP (in sulfur dioxide equivalent) addresses the impacts from acidification generated by the emission of airborne acidifying chemicals. Acidification refers literally to processes that increase the acidity of water and soil systems by hydrogen ion concentration (Institute for Environment and Sustainability of Joint Research Centre of European Commission 2010). Then there's POCP (in kg ethene equivalent), which rates the creation of ozone (due to reaction of a substance in presence of NO_x gases), also known as summer smog (Guinée et al. 2002). The negative impact causes respiratory diseases and oxidative damage on photosynthetic organelles in plants (Institute for Environment and Sustainability of Joint Research Centre of European Commission 2010). In the final phase, the results from life cycle impact assessments are interpreted.

2.2   Social Assessment

In the SLCA guidelines, the methodology framework is proposed similar to LCA: goal and scope definition, life cycle inventory analysis, life cycle impact assessment and interpretation (Chang et al. 2012; UNEP 2009). In the guidelines, five main stakeholder groups (workers, consumers, local community, society and value chain actors) and 31 subcategories are described and the relevant social issues are then listed. Due to the high level of importance held by the stakeholders responsible for welders' welfare in Germany, the two critical social conditions "fair salary" and "health and safety" have been selected for the social assessment.

The sufficiency status of salary for welders in Germany can be recognized by comparing the average wage of welders (FOCUS Online 2012), the national minimum wage (Statistisches Bundesamt 2016a) and at-risk-of-poverty threshold (Statistisches Bundesamt 2016b). The at-risk-of-poverty threshold serves as a yardstick for identifying whether people live in income-dependent poverty. In this chapter, gross monthly wage and poverty threshold for a single person are used for comparison. The reference year for the national minimum wage and at-risk-of-poverty threshold is 2015, but the average wage of welders is taken from 2011 due to the statistical data limitation.

In addition to fair salary, the relative health and safety effects on welders performing different welding technologies have been analysed, with a specific look at exposure to welding fumes. Welding processes generate a complex mixture of fumes (respirable and ultrafine particles) as by-products composed of an array of metals volatilised from the welding electrode or the flux materials incorporated (Antonini et al. 2006). Welders' exposure to welding fumes is often associated with acute and chronic lung damage, lung cancer and other potential harm on heart, kidneys and central nervous systems (Gonser and Hogan 2011; Canadian Centre for Occupational Health & Safety 2016). Iron oxides constitute the main part of the fume, while chromium, manganese, and nickel account for the total remaining fume composition (Antonini et al. 2006; Jenkins and Eagar 2005). Iron oxide is not officially classified as a human carcinogen. Nevertheless, it has proven to trigger siderosis, which decreases lung capacity. Chromium (VI, insoluble) and its compounds are known as a human lung carcinogen, while nickel is also known as a human carcinogen, causing lung, nasal, and sinus cancers. Manganese and its compounds are not carcinogens, but associated with central nervous system (CNS) effects similar in nature to Parkinsonism (Gonser and Hogan 2011). To represent the relative potential risk caused by fumes on the health of welders, we have identified the hazard figure (Gefährdungszahl, GZ) of the welding processes. Based on the literature (Spiegel-Ciobanu 2012), the model simplifies and considers process-specific fume emissions associated with the working situation. For estimating the simplified potential risk GZ_s, the following Eq. 1 is used (Spiegel-Ciobanu 2012):

$$GZ_s = \left(E_p \times W_p\right) \times L \times R \times K_b \qquad (1)$$

E_p = emission factor of the specific substance per functional unit;
W_p = potential effect for the specific substances in fume;
L = ventilation factor (have sufficient ventilation or not);
R = spatial factor (outside or in rooms);
K_b = the factor of relative distance of head/body and fume source.

E_p represents the fume emissions per functional unit of 1 m weld seam and is calculated based on the inventory data for fume emissions of the LCA (see Sect. 2.1). It is a relative factor taking the minimal emissions per functional unit as a reference value. Since the distance between welders and fume sources in different welding processes vary widely, the K_b levels are set correspondingly. The closer distance indicates a higher chance of inhaling fume. MMAW is executed manually and welders are close to the fume sources, so the levels are set as 4 (Spiegel-Ciobanu 2012); in GMAW (executed with a robot), the K_b level is assumed as 2 due to there usually being some distance between welders and fume sources; in LAHW, the welding process is performed in welding cells, so the K_b level is defined as 1 (Spiegel-Ciobanu 2012). Targeting the comparison of potential risks, W_p can be assumed as the same value as 1 to represent no difference in comparison between the processes since the composition of materials in the chosen welding processes are highly similar. Also, the L and R both are set as 1 in the paper due to the condition of welding places assumed to be identical. Following Eq. 1 and the assumptions, the potential health risk, GZ_s, is highly influenced by the emission factor per functional unit E_p and the relative distance of head/body and fume source K_b. The GZ_s can be simply represented as $E_p \times K_b$.

2.3 Welding Experiments

Welding was carried out in four types of technologies: MMAW, LAHW, and GMAW in modified spray arc (GMAW modified) and the conventional spray arc (GMAW standard). Low alloyed structural steels and proper filler wires were used as a base and filler metal. Weld samples were plates of 20 mm thickness with weld seam lengths from 250 to 300 mm. Welding was performed in the flat position

Table 2 Material, joint specifications and process parameters of MMAW and LAHW

	MMAW	LAHW
Joint preparation	Double-V (ISO 9692-1) 60° groove angle 2 mm root gap 2 mm root face	Y-groove (ISO 9692-1) 45° groove angle No root gap 14 mm root face
Base material	S355 + N (DIN EN 10025-3)	X120 (API 5L)
Filler material	E 42 0 RR 1 2 (DIN EN ISO 2560-A)	Mn4Ni2CrMo (DIN EN ISO 16834)
Shielding gas	–	82 % Argon, 18 % CO_2
Process parameters		
Average welding speed in mm/s	2.8	LAHW: 43.3 GMAW filler pass: 13.3
Number of passes	8	2
Average power in kW	4	Root pass: 33 (Laser + GMAW) Filler pass: 11 (GMAW only)

Table 3 Material, joint specifications and process parameters of GMAW standard and GMAW modified

	GMAW standard	GMAW modified
Joint preparation	Double-V (ISO 9692-1), 60° groove angle 0.4 mm root gap 2 mm root face	Double-V (ISO 9692-1) 30° groove angle 0.2 mm root gap 2 mm root face
Base material	S690 QL (DIN EN 10025-6)	S960 QL (DIN EN 10025-6)
Filler material	Mn3Ni1CrMo (DIN EN ISO 16834)	Mn4Ni2CrMo (DIN EN ISO 16834)
Shielding gas	82 % Argon, 18 % CO_2	
Process parameters		
Average welding speed in mm/s	6.2	6.7
Number of passes	4	2
Average power in kW	8	12

(1 G) and data was calculated with regards to the functional unit of 1 m weld seam. Material specifications, groove preparations and process parameters of the processes are listed in Tables 2 and 3.

3 Case Study Results and Discussion

3.1 Environmental Assessment

The LCA study highlights the environmental impacts contributed by different inputs and outputs of the chosen welding processes and compares the differences of environmental impacts. The life cycle inventory data is shown in Table 4 based on the functional unit. The inventory is used to conduct life cycle impact assessments.

By carrying out impact assessment within CML method and GaBi 6.0 software, the environmental impacts GWP, EP, AP and POCP contributed by the selected welding processes have been estimated, as shown in Fig. 2. The results indicate that MMAW causes the highest environmental impact in the chosen impact categories among the selected processes, and the LAHW variant provides the lowest. In addition, the modified spray arc with the smaller groove angle contributes significantly lower impact than the standard GMAW variant. For GMAW and LAHW, electric energy and filler material are the dominant influencing factors. For MMAW, the electrode coating is of major relevance, along with filler material and electric energy.

Table 4 Life cycle inventory of the welding processes

	MMAW	GMAW standard	GMAW modified	LAHW
Filler material in g	944	890	530	155
Shielding gas in l	–	241	100	33
Electrode coating in g	580	–	–	–
Compressed air for laser optics cross-jet in l	–	–	–	249
Electric energy in kWh	3.9	2.1	1.3	0.9
Welding fumes in g	11.6	3.6	1.2	0.6
Slag in g	600	–	–	–
Electrode stubs in g	150	–	–	–

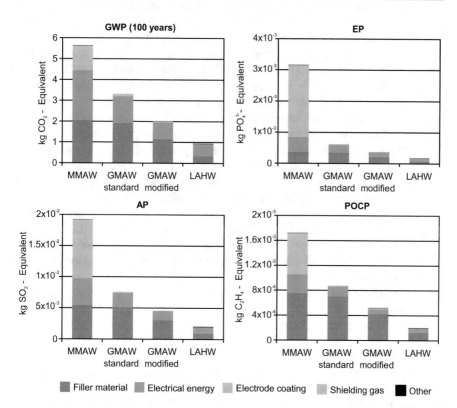

Fig. 2 Results of the impact assessment

Among the processes investigated in joining a 20 mm thick plate of structural steels, LAHW is the best option hands down when considering the environmental impact caused. Due to its high power density, LAHW performs welding with both the least number of passes and overall weld volume. Additionally, LAHW allows for high welding speed, leading to high productivity and low electricity and gas consumptions. This is a remarkable finding considering the low beam source efficiency of 30 % in contrast to 80 % efficiency of arc welding machines. The main reason for less environmental impact in LAHW lies in the better ratio between power consumed and welding time, which means that the low efficiency is over-compensated by welding time savings. Either filler material or electric energy is dominant depending on the indicator considered. Both can be optimized by means of enlargement of the root face width and a smaller opening angle. Moreover, electric energy consumption could be further reduced significantly by increasing the beam source efficiency.

Contrary to LAHW, low process performance (deposition rate and welding speed) and the necessary edge preparation in MMAW lead to the highest environmental effects. Low deposition rate and welding speed result in higher amounts of energy that are used to re-melt weld metal in the subsequent passes, as well as energy losses due to heat conduction into the base material. Furthermore, electrode coating accounts for a remarkable share of environmental impact even though only 55 % of the electrode composition is considered in the LCA model. It is likely that results would be even worse for MMAW if electrode coating could be fully accounted for. In order to mitigate environmental impact, the industry should therefore focus firstly on rutile electrode coatings and then on joint design. Smaller root gaps and opening angles would reduce electric energy and material (filler as well as coatings) consumption. Thickness of electrode coatings can be reduced and alternative compositions can be further investigated (e.g. basic or acid coated electrodes) with respect to their environmental impact.

Filler material consumption dominates about 54–80 % of the instances of impact in GMAW in the chosen categories. The benefit of reducing opening angles can be directly stated by comparing GMAW with the standard spray arc and the modified spray arc. This leads to approximately 40 % reduction of the environmental impact level. Hence, in order to improve GMAW from an environmental perspective, joints should always be designed with the minimum possible flange angle. However, it is unclear whether optimisation options are technologically feasible or whether they guarantee the optimal weld performance, all of which should be evaluated properly.

Welding robot movements for all technologies account for a small share of electricity consumption. As a result, the energy efficiency attached to joining industrial parts is dominated by the welding process itself and has to be adequately assessed in future work accordingly.

The LCA results show clear environmental preferences. Nevertheless, gaps and limitations of the study must be acknowledged, for example the challenges embedded in LCA methodology (Finkbeiner et al. 2014) and the possible variation of results due to different process requirements in welding technology. Process requirements such as efforts for edge preparation, effects of different welding positions or mobility of equipment could furthermore have a crucial influence on process selection. In the LCA model, only four impact categories are considered for comparison, which can lead to inconclusive judgment. What's more, machinery is not considered, which could cause potential bias and require a critical overall weld seam length before proving to be environmentally beneficial.

3.2 Social Assessment

The latest salary survey from Focus Online (FOCUS Online 2012) showed the average gross salary per month of welders in Germany in 2011 to be €2,165; the national minimum wage was €1,430 (Statistisches Bundesamt 2016a); and the poverty threshold for a single person was deduced to be €986.67 based on the national statistics (Statistisches Bundesamt 2016b). The results indicate that the average monthly wage of welders is higher than the current national minimum wage and the deduced poverty threshold (approximate 2 times). It is therefore fair to conclude that welders' salary status is sufficient for supporting their overall subsistence and for meeting the income regulation of minimum wage.

The evaluation of the potential risks GZ_s of the applied welding processes are displayed in Table 5. The emission factors E_p are calculated based on the inventory data shown in Table 4, taking the emission of LAHW as the reference value for estimating the ratios. Thus, LAHW constitutes the lowest potential health risk. This is because it is conducted in closed cells due to laser safety restrictions. MMAW owns the highest GZ_s to welders among all the selected processes. GMAW standard and GMAW modified have smaller differences of the GZ_s since they only differ in the quantity of fume formation. The results underline that welders working in the manual processes (like MMAW) face higher risks than in automatic processes (GMAW, LAHW). Consequently, it is important to limit the application of manual welding processes to the minimum possible extent. Moreover, the personal protective equipment used should be adequate to minimize the health risks for welders. In case of automatic GMAW, the future goal should be to keep welders out of the process zone. However, this requires technologies for advanced process control and monitoring to ensure the quality of the welds. Apart from the potential health risks posed by welding fumes, further factors in welding contribute to the category "health and safety." In particular, electrical, thermal and radiation hazards or the workplace ergonomics should be evaluated in the future in the pursuit of an improved working environment for welders.

In summary, the SLCA showed a sufficient wage level from which welders may support themselves financially. Potential health risks of operation depend on the respective process and are high for manual processes such as MMAW.

	E_p	K_b	GZ_s
MMAW	19	4	76
GMAW standard	6	2	12
GMAW modified	2	2	4
LAHW	1	1	1

Table 5 The estimation of relative health effects of the welding processes

4 Conclusions

This contribution evaluates the environmental impact and social influences of welding technologies by applying LCA and SLCA. It provides information to the industry as well as to the research community for developing and selecting joining technologies in view of the triple bottom line of sustainability.

The instances of environmental impact involved the selected impact categories of eutrophication potential, acidification potential, global warming potential (100 years) and photochemical ozone creation potential. The social categories were "fair salary" and "health and safety." The results serve to support industry in the development and selection of sustainable joining technologies.

The LCA results show that MMAW contributes higher environmental impact levels than GMAW or LAHW. The main cause is that MMAW consumes much more material and electricity per 1 m weld seam. Titanium dioxide consumption for electrode coating in MMAW is critical in contributing the main burden of acidification and eutrophication. GMAW is strongly influenced by filler material consumption, which is governed by the seam preparation. This is improved by using a modified spray arc, which ultimately enables a reduction of flange angles from 60° to 30°. Within the scope of the study, LAHW stands as the superior technology.

The social LCA revealed a sufficient salary for welders and potential health risks that depend on the applied process. LAHW demonstrates the lowest and MMAW the highest potential health risks that arise from fume formation. Especially manual technologies such as MMAW should therefore be limited to the minimum possible extent to reduce health risks for welders.

References

Antonini, J.M., A.A. Afshari, S. Stone, B. Chen, D. Schwegler-Berry, W.G. Fletcher, W.T. Goldsmith, et al. 2006. Design, construction, and characterization of a novel robotic welding fume generator and inhalation exposure system for laboratory animals. *Journal of Occupational and Environmental Hygiene* 3(4): 194–203. doi:10.1080/15459620600584352.

Canadian Centre for Occupational Health & Safety. 2016. Occupations & workplaces: welders. Canadian Centre for Occupational Health & Safety. http://www.ccohs.ca/oshanswers/occup_workplace/welder.html. Accessed 25 July 2016.

Chang, Y.-J., E. Schau, and M. Finkbeiner. 2012. Application of life cycle sustainability assessment to the bamboo and aluminum bicycle in surveying social risks of developing countries. Paper presented at the 2nd World Sustainability Forum, Web Conference, 1–30 November 2012.

Dzelnitzki, D. 2000. Vergrößerung des Einbringvolumens oder Erhöhung der Schweißgesch windigkeit ? –Vorteile des MAG-Hochleistungsschweißens. https://www.ewm-group.com/de/service/downloads/fachaufsaetze/migmag-schweissen/136-vergroesserung-des-einbringvolumens-oder-erhoehung-der-schweissgeschwindigkeit-vorteile-des-mag-hochleistungsschweissens.html. Accessed 25 July 2016.

Finkbeiner, M., R. Ackermann, V. Bach, M. Berger, G. Brankatschk, Y.-J. Chang, M. Grinberg, et al. 2014. Challenges in life cycle assessment: An overview of current gaps and research needs. In *Background and future prospects in life cycle assessment*, ed. Walter Klöpffer, 207–258. Dordrecht: Springer.

FOCUS Online. 2012. 150 Berufe im Gehaltsüberblick. http://www.focus.de/finanzen/karriere/tid-28806/titel-deutschlands-groesster-gehalts-report-150-berufe-im-gehaltsueberblick_aid_89032 8.html. Accessed 2 May 2014.

Gonser, M. and T. Hogan. 2011. Arc welding health effects, fume formation mechanisms, and characterization methods. In *Arc welding*, ed. by Wladislav Sudnik. doi:10.5772/29131.

Guinée, J.B., M. Gorrée, R. Heijungs, G. Huppes, R. Kleijn, A. de Koning, L. van Oers, et al. 2002. *Handbook on life cycle assessment—operational guide to the ISO standards*. New York: Kluwer Academic Publishers.

Hälsig, A. 2014. Energetische Bilanzierung von Lichtbogenschweißverfahren. Ph.D. dissertation, Technische Universität Chemnitz.

Institute for Environment and Sustainability of Joint Research Centre of European Commission. 2010. *ILCD handbook: Framework and requirements for LCIA models and indicators*. Luxembourg: Publications Office of the European Union.

ISO. 2006a. *ISO 14040—environmental management—life cycle assessment—principles and framework (ISO 14040:2006)*. Geneva: ISO—International Organization for Standardization.

ISO. 2006b. *ISO 14044—environmental management—life cycle assessment—requirements and guidelines*. Geneva: ISO—International Organization for Standardisation.

Jenkins, N.T., and T.W. Eagar. 2005. Chemical analysis of welding fume particles. *Welding Journal* 84(6): 87s–93s.

Klöpffer, W., and B. Grahl. 2009. *Ökobilanz (LCA): ein Leitfaden für Ausbildung und Beruf*. Weinheim: Wiley-VCH Verlag.

Lezzi, F., and L. Costa. 2013. The development of conventional welding processes in naval construction. *Welding International* 27(10): 786–797. doi:10.1080/09507116.2012.753256.

Neugebauer, S., J. Martinez-Blanco, R. Scheumann, and M. Finkbeiner. 2015. Enhancing the practical implementation of life cycle sustainability assessment—proposal of a Tiered approach. *Journal of Cleaner Production* 102: 165–176. doi:10.1016/j.jclepro.2015.04.053.

PE Interational. 2014. Harmonization of LCA methodologies for metals—a whitepaper providing guidance for conducting LCAs for metals and metal products. https://www.icmm.com/document/6657. Accessed 9 July 2016.

Pohlmann, G., C. Holzinger, and V.E. Spiegel-Ciobanu. 2013. Comparative investigations in order to characterize ultrafine particles in fumes in the case of welding and allied processes. *Welding and Cutting* 12(2): 97–105.

Rose, S., M. Schnick, and U. Füssel. 2012. Cause-and-effect chains of fume formation in GMAW and possibilities of fume reduction by using new welding processes. IIW Document VIII-2147-12.

Schau, E.M., Y.-J. Chang, R. Scheumann, and M. Finkbeiner. 2012. Manufactured products—how can their life cycle sustainability be measured?—A case study of a bamboo bicycle. Paper presented at The 10th Global Conference on Sustainable Manufacturing, October 31–November 2, Istanbul, Turkey.

Spiegel-Ciobanu, V.E. 2012. Occupational health and safety regulations with regard to welding and assessment of the exposure to welding fumes and of their effect. *Welding and Cutting* 11 (1).

Sproesser, G., Y. Chang, A. Pittner, M. Finkbeiner, and M. Rethmeier. 2015. Life cycle assessment of welding technologies for thick metal plate welds. *Journal of Cleaner Production* 108: 46–53. doi:10.1016/j.jclepro.2015.06.121.

Sproesser, G., A. Pittner, and M. Rethmeier. 2016. Increasing performance and energy efficiency of gas metal arc welding by a high power tandem process. *Procedia CIRP* 40(2016): 642–647. doi:10.1016/j.procir.2016.01.148.

Statistisches Bundesamt. 2016a. Minimum wages. https://www.destatis.de/EN/FactsFigures/NationalEconomyEnvironment/EarningsLabourCosts/MinimumWages/MinimumWages.html. Accessed 25 July 2016.

Statistisches Bundesamt. 2016b. Poverty threshold and risk of poverty in Germany. https://www.destatis.de/EN/FactsFigures/SocietyState/IncomeConsumptionLivingConditions/LivingConditionsRiskPoverty/Tables/EU_PovertyThresholdRisk_SILC.html. Accessed 25 July 2016.

UNEP. 2009. The guidelines for social life cycle assessment of products. Nairobi, Kenya: United Nations Environment Programme. http://www.unep.org/pdf/DTIE_PDFS/DTIx1164xPA-guidelines_sLCA.pdf.

World Steel Association. 2011. *Life cycle assessment methodology report*. Brussels: World Steel Association.

Human-Centred Automation to Simplify the Path to Social and Economic Sustainability

The Duy Nguyen and Jörg Krüger

Abstract Musculoskeletal Disorders (MSDs) pose a serious threat to sustainability in manufacturing. In particular, this phenomenon impacts the sustainability indicators of worker health and safety and the Gross Domestic Product (GDP). Effective MSD prevention measures would therefore constitute a remarkable contribution to social and economic sustainability. This chapter provides first an outline of existing methods to prevent MSD at the workplace. Analysis of the approaches yields that effective solutions require earmarked finances as well as qualified personnel, both of which are not affordable for many companies. In pursuit of solutions, Human-centred Automation (HCA), a recent paradigm in manufacturing, proposes the design of manufacturing systems using intelligent technology to support the worker instead of replacing him/her. HCA has the unique potential of reducing the effort needed to implement MSD prevention strategies by simplifying the path to social and economic sustainability. This chapter demonstrates this process with the example of the "Working Posture Controller" (WPC), which illustrates how the HCA concept can be applied. Finally, the lessons learned from the case are outlined, providing a vision of how future workplaces can benefit from HCA.

Keywords Musculoskeletal disorders · Human-centred automation · Human-machine interaction

1 Work-Related Musculoskeletal Disorders—A Sustainability Challenge

The health of the workforce is vital for social as well as economic sustainability. The Guideline for Social Life Cycle Assessment of Products (Benoît et al. 2010) describes "worker health and safety" as a major impact category among social

T.D. Nguyen (✉) · J. Krüger
Technische Universität Berlin, Berlin, Germany
e-mail: theduy.nguyen@iat.tu-berlin.de

© The Author(s) 2017
R. Stark et al. (eds.), *Sustainable Manufacturing*, Sustainable Production,
Life Cycle Engineering and Management, DOI 10.1007/978-3-319-48514-0_6

sustainability indicators. In terms of economic sustainability, direct costs due to unfavourable working conditions reduce a country's Gross Domestic Product (GDP) (Bevan 2015), which is considered to be one of the main economic sustainability indicators defined by the United Nations—Department of Economic and Social Affairs (2007).

Musculoskeletal Disorders (MSDs) present a serious threat to the health of the workforce, and thus, to sustainability. The European Agency for Safety and Health at Work (Schneider et al. 2010) defines MSDs as "health problems of the locomotor apparatus, which includes muscles, tendons, the skeleton, cartilage, the vascular system, ligaments and nerves." Work-related or Occupational Musculoskeletal Disorders (WMSDs) encompass all MSDs that are caused or worsened by work.

WMSDs as a sustainability indicator are not explicitly mentioned in the "health and safety" impact group in Benoît et al. (2010). However, this represents more of an oversimplification of the guideline than a negligible effect. In fact, the only measures which are considered are those which result from suboptimal working conditions, such as the number of injuries or accidents (Chang et al. 2016). Accumulative effects, such as WMSDs go completely neglected although they pose a comparable impact. The European Labour Force Survey (Camarota 2007) concluded that MSDs accounted for 53 % of all work-related diseases in the EU-15, therein representing the most frequent cause (Bevan 2015). The number of lost days due to WMSDs is estimated at 350 million (Delleman et al. 2004) in the EU. In terms of economic sustainability, WMSDs significantly reduce the GDP of the EU. The total costs of WMSDs is estimated at 240€ billion, which translates into up to 2 % of the EU GDP (Bevan 2015).

Due to its impact, researchers from different scientific disciplines, such as human factors science, medicine and engineering, have developed methods to prevent WMSDs, reducing their risk of occurrence. Significant successes have been achieved. On average, methods implemented have turned out to cover their total costs in less than 1 year (Goggins et al. 2008). Nevertheless, implementing effective measures requires tedious work on the part of highly qualified ergonomists, which makes effective WMSD prevention not realisable for every company.

Human Centred Automation (HCA) denotes a recent development in manufacturing technology. This engineering paradigm proposes turning away from fully automated production lines in favour of systems where man and machine collaborate and combine the strengths of both participants. Instead of replacing the worker, the machine's task is to support him/her. The system can enhance cognitive skills through intelligent sensors or provide additional physical capabilities through actuators. By automatising the parts of the WMSD prevention methods which require highly skilled personnel or tedious work, HCA helps to make these techniques available to a broader mass. To sum up, the contribution of HCA to sustainability lies in providing access to sustainability enhancing techniques.

This chapter concentrates on one main risk factor causing WMSDs: unfavourable working posture, which is often referred to in literature as "awkward posture" (Delleman et al. 2004). An exemplary technique is presented on how HCA can be applied to solve existing WMSD prevention problems, and thus, support sustainability goals in manufacturing. Section 2 provides an overview of common state-of-the-art approaches in tackling WMSDs, outlining a fundamental problem: the effectiveness–flexibility trade-off. The HCA, which appears to be a promising solution to the effectiveness–flexibility trade-off, is presented in Sect. 3. Afterwards, Sect. 4 presents the Working Posture Controller (WPC). The WPC is a device which demonstrates how the HCA paradigm is used to overcome the effectiveness–flexibility trade-off. Finally, this chapter concludes with the facts learned.

2 State-of-the-Art of WMSD Prevention

Due to its high impact on human health and the economy, the area of WMSD prevention is an extensive research field. Researchers from various disciplines such as, human factors, medicine or engineering, have proposed their solutions. In scientific literature, the measures are often referred to as "ergonomic interventions." This section outlines the most important developments.

In brief, the techniques presented can be grouped into three categories: technical measures, organisational measures or individual measures (Van der Molen et al. 2005). Alternatively, Bergamasco et al. (1998) use the term "training" instead of individual measure.

Technical measures involve modifications of the working environment and the process. Examples include designing the workplace layout, process design, or the introduction of special equipment to support the worker. Workplace layout design aims at rearranging the workplace geometry in such a way that tasks can be efficiently accomplished without the need for adopting awkward postures. To that effect, ergonomic guidelines have been released to provide the workplace designer with a tool for checking the appropriateness of the developed workplace (Das and Grady 1983). The set of ergonomic guidelines is complex and highly dependent on the tasks at hand and on the individual person. Often, multiple physical prototypes have to be evaluated (Delleman et al. 2004). Digital Human Models (DHM) have become a popular method for assisting in the design process (Lämkull et al. 2009) by means of simulating the prototypes. Technical measures also imply the introduction of equipment, such as lifting aids or human robot collaboration systems to execute physically demanding tasks on behalf of the worker (Krüger et al. 2009; Busch et al. 2012; Weidner et al. 2013; Schmidtler et al. 2014). Another type of equipment is found in alert systems which monitor the process and warn the user as soon as an ergonomically unfavourable situation arises (Vignais et al. 2013).

In addition, organisational measures (Paul et al. 1999) entail techniques which aim at avoiding an inacceptable amount of load through customised and calculated, balanced scheduling of the tasks. The idea is to compose the set of tasks for a worker in a way such that multiple regions of the body are alternatingly strained. This avoids the monotonous strain of one particular body structure leading to long-term damage. This technique is called job rotation.

Finally, individual measures (Engels et al. 1998) supply the worker with basic knowledge about best practices so as to enable the preservation of one's own health. This can include biomechanical theory, general ergonomics, or techniques e.g. for lifting. Often, the theoretical courses are complemented with practical training.

Multiple approaches are available for tackling the WMSD dilemma. A decision-maker has the challenge of selecting the most promising approaches to be implemented. To that end, Goggins et al. (2008) have proposed a scale for ranking the measures according to their effectiveness (see Fig. 1). The scale is based on a Cost-Benefit analysis derived from the review of around 250 studies in industrial as well as office environment. This scale comprises four classes:

- measures that completely eliminate the exposure (level 1)
- measures that reduce the level of exposure (level 2)
- methods that reduce the exposure time (level 3)
- and methods, whose success merely rely on the worker's behaviour (level 4).

Level 1 methods are the most effective ones whereas level 4 methods should only be considered if the other measures are infeasible.

Apart from their effectiveness, the specific amount of effort required to implement the measures has to be considered. The techniques can be grouped into two categories: pre-process techniques and in-process techniques. Pre-process techniques require the effort attached to tailoring the method to the individual task and user groups at hand. In changing the production environment requiring flexible production systems, these measures can produce a bottleneck. Examples are workplace design or organizational measures. In-process measures, on the other hand, only require one initial setup routine and then adapt to changing situations. Examples are found in so-called *cobots* or alert systems. Their impact is however limited, since their adaptation normally only covers few well-defined cases. In conclusion, a fundamental trade-off exists between the effectiveness and the flexibility of the ergonomic intervention techniques.

Though entailing tremendous effort, Goggins et al. (2008) state that the impact of the interventions measured has been highly promising. Incidence rates have, for example, dropped by 65 % on average and companies had to pay 68 % of the original compensation costs after intervention. Additionally, the productivity of the workforce and the quality of the products have improved. Most notably, these

Fig. 1 Scale of effectiveness according to Goggins et al. (2008)

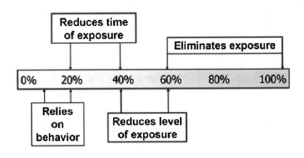

effects appear within less than 1 year after implementation. The challenge lies in addressing how to achieve a high effect without losing out on flexibility.

3   The Potential of Human-Centred Automation (HCA) for Sustainability

With the introduction of computers, manufacturing has been veritably revolutionised. Tasks which had been time-consuming and tedious can now be efficiently performed with relative ease. At the same time, tasks which originally required human experts have been simplified in such a way that lower skilled operators can use them.

In recent years, techniques of Human-centred Automation (HCA) have become a matter of research. The original term comes from the domain of aviation and was introduced by Bilings (1997). HCA describes a novel approach in system design. It proposes building an environment, "in which humans and machines collaborate cooperatively in order to reach stated objectives." This paradigm has been applied in various systems, such as driver assistance systems, aircraft flight management systems and air traffic systems.

Furthermore, HCA has increasingly become a matter of research in manufacturing. In this domain, HCA represents a new paradigm of turning away from full automatisation as a long-term goal, and instead moves in the direction of achieving more flexible manufacturing structures. Systems of HCA strive to support the worker rather than to replace him/her with technology. The human remains the core of the process and the technology is used to enhance cognitive skills by sensors and physical skills by actuators. The result is a production system wherein worker and machine are tightly bound together, combining one other's strengths and compensating for each other's weakness.

To be sure, both human and automation systems have their advantages and shortcomings. Humans are highly flexible insofar as being able to learn new tasks in a short time. Yet, the human is vulnerable. Automation technology with its actuators proves however to be quite powerful and can efficiently accomplish repetitive tasks. On the other hand, flexibility is lost in that process, since programming the machines likewise takes time and is work intensive, making it only suitable in high lot size production scenarios.

Recent developments in the field of intelligent systems have made it possible to develop systems which support the human in a more sophisticated manner. The trouble with the solutions mentioned in Sect. 2 is that they require human expertise. Considering the lack of experts to implement the solutions, the only way to escape this dilemma is to teach machines to take over some of the tedious as well as sophisticated manual tasks. In this way, HCA can help to implement measures designed to reduce health risks at the workplace with an acceptable level of effort.

4   The Working Posture Controller (WPC)

WMSDs present a grave problem for the manufacturing community. Especially considering the ageing worker population in many countries, these disorders are poised to become a significant sustainability challenge. Remarkable effort has been put into solving the WMSD issue. Yet, a fundamental trade-off in the solutions proposed stands at the crossroads: either the measures are effective but inflexible, or they are easily adaptable but less effective. Nevertheless, some experts have concluded that HCA techniques are starting to make an impact on manufacturing by combining the best of both worlds. A particular motivation for HCA has been the need for flexibility without the need to relinquish the advantages bestowed by automation technology.

All these facts taken together indicate that the solution to the problem seems to lie in applying HCA techniques to confront the WMSD challenge. The questions to be posed are: What exactly do the resulting systems look like? How can such systems be technically realised? This section attempts to provide answers by presenting an exemplary device: the Working Posture Controller (WPC) (Krüger and Nguyen 2015; Nguyen et al. 2016).

The WPC is a system that continuously monitors the worker's posture in the process and adjusts the workplace layout when the combination of task and workplace does not allow for a natural working posture. Through automatising the work and knowledge-intensive parts in the highly effective measures, flexibility is gained.

4.1   Concept of the WPC

The main conceptual idea of the WPC is to combine automatised ergonomic assessments with automatised workplace design into one system. This system enables the user to avoid adopting awkward posture for prolonged periods. In the workflow, man and machine are embedded into a control loop. Figure 2 depicts one full sequence. A posture assessment module monitors the worker's posture, assigning a numerical score which represents the current postural load of the task. If the score exceeds a particular value, the system initiates the posture optimisation module. This component first interprets the worker's posture to interpret which task he/she intends to accomplish. Afterwards, it computes a workplace layout adjustment which enables accomplishing the same task with a more ergonomic posture. Having found this adjustment, the system then gives feedback to the user, providing a visual of the proposed workplace geometry and posture. Once the worker agrees, the adjustment is executed. Upon adjustment, the cycle then starts anew. The coming subsections describe the two components in further detail.

Fig. 2 Concept of the WPC. An actuator, in this case a robot, holds the workpiece to be processed. Additionally, a sensor system monitors the worker's posture. In case the posture becomes physically straining, the WPC proposes the change to the workpiece pose, making it possible for the user to adopt a more natural posture

4.2 Posture Assessment

The posture assessment module is based on the "Ergonomic Assessment Worksheet (EAWS)" (Schaub et al. 2013), a manual tool to help ergonomic practitioners assessing working tasks. The whole concept behind it is that the practitioner observes the task and assigns load points for the duration of the particular working postures adopted. The EAWS defines a set of postures which have to be recognised from observation. The accumulated score of the single posture durations yields the overall postural risk score.

Automatising this posture assessment process requires the system to automatically recognise the right posture from the defined one. The WPC uses a consumer depth camera to acquire the input data. Afterwards, a markerless motion capture algorithm (Nguyen et al. 2014) is developed to determine the coordinates of each limb. Having obtained the coordinates, classifiers are trained on training datasets to recognise the posture. Upon recognition, load points can be assigned and the current risk score can be accumulated accordingly. If this ergonomic score exceeds a given threshold, an adjustment is then initiated.

4.3 Posture Optimisation

Having detected the ergonomically critical situation, the posture optimisation algorithm attempts to compute an alternative workplace adjustment where the worker is able to accomplish the same task in a healthier posture.

The algorithm first interprets the original task intended from the recognised posture. Relevant parameters to be detected lie in the orientation of the upper arm and the location of the hands relative to the workpiece. Afterwards, the algorithm searches for an adjustment of the workplace geometry, which enables accomplishing the task in an ergonomically more favourable posture. The biggest challenge lies in mathematically modelling the human behaviour at hand. The posture

Fig. 3 The *red manikin*
denotes the current posture
whereas the *green manikin*
denotes the optimised
working posture after
adjustment

adopted once a potential adjustment has been taken, then has to be predicted. The behaviour is modelled by transforming the task into non-linear mathematical optimisation problem which can be solved with standard optimisation algorithms. Different types of actuators yield different optimisation problems. An exemplary visualisation of original and predicted posture after adjustment is depicted in Fig. 3.

5   Discussion and Outlook

WMSDs have a high negative impact on social as well as economic sustainability. The WPC shows that HCA can be successfully applied to improve WMSD prevention methods. This novel way of preventing awkward posture combines sensors, intelligent algorithms and actuators to enable the machine to perform tasks which would normally require human expertise in less time. Through the time-intensive and tedious process of setting up the parts of the workplace design to fit the automation technology, the whole production system gains the flexibility required. The system is designed to be usable with cost-efficient as well as advanced hardware in order to address a broad group of users.

Implementing HCA concepts into further manufacturing requires a discussion of the fundamental problems of man-machine interaction. First of all, engineers have to define the appropriate level of automation (LoA) for the given system. The term LoA is defined by Frohm (2008) as "allocation of physical and cognitive tasks between humans and technology, described as a continuum ranging from totally manual to total automation." The problem is summed up by the question: what is supposed to be done by the human and what is supposed to be done by technology? The answer to this question will influence what such systems look like, and in what manner they assist the worker. Second, after defining how to allocate the task, designers have to define how human and technology are supposed to communicate. This aspect is critical for the acceptance of HCA. Common mistakes in the design of communication between human and machine are described in such works as Endsley's (1995).

To conclude, HCA stands as a promising means of supporting social and economic sustainability goals. However, designing these systems is challenging, since it is not known exactly what form they ultimately take. The WPC, among other projects, has shown one example of how such a system can be designed.

References

Benoît, C., G.A. Norris, S. Valdivia, A. Ciroth, A. Moberg, U. Bos, S. Prakash, C. Ugaya, and T. Beck. 2010. The guidelines for social life cycle assessment of products: Just in time! *The International Journal of Life Cycle Assessment* 15(2): 156–163. doi:10.1007/s11367-009-0147-8.

Bergamasco, R., C. Girola, and D. Colombini. 1998. Guidelines for designing jobs featuring repetitive tasks. *Ergonomics* 41(9): 1364–1383. doi:10.1080/001401398186379.

Bevan, S. 2015. Economic impact of musculoskeletal disorders (MSDs) on work in Europe. *Best Practice & Research Clinical Rheumatology* 29(3): 356–373. doi:10.1016/j.berh.2015.08.002.

Billings, C.E. 1997. *Aviation automation: The search for a human-centered approach*. Abingdon: Taylor & Francis.

Busch, F., J. Deuse, and B. Kuhlenkötter. 2012. A hybrid human-robot assistance system for welding operations—methods to ensure process quality and forecast ergonomic conditions. In *Technologies and Systems for Assembly Quality, Productivity and Customization—Proceedings of 4th CIRP Conference on Assembly Technologies and Systems (CATS)*, ed. S. J. Hu, 20–22. Ann Arbor: University of Michigan.

Cammarota, A. 2007. *The European Commission initiative on WRMSDs: Recent developments*. Presentation at EUROFOUND conference on 'Musculoskeletal disorders', Lisbon, October 11–12.

Chang, Y.J., T.D. Nguyen, M. Finkbeiner, and J. Krüger. 2016. Adapting ergonomic assessments to social life cycle assessment. *Procedia CIRP* 40: 91–96. doi:10.1016/j.procir.2016.01.064.

Das, B., and R.M. Grady. 1983. Industrial workplace layout design an application of engineering anthropometry. *Ergonomics* 26(5): 433–447. doi:10.1080/00140138308963360.

Delleman, N.J., C.M. Haslegrave, and D.B. Chaffin. 2004. *Working postures and movements: Tools for evaluation and engineering*. Boca Raton: CRC Press.

Endsley, M.R., and E.O. Kiris. 1995. The out-of-the-loop performance problem and level of control in automation. *Human Factors: The Journal of the Human Factors and Ergonomics Society* 37(2): 381–394.

Engels, J.A., J.W.J. Van der Gulden, T.F. Senden, J.J. Kolk, and R.A. Binkhorst. 1998. The effects of an ergonomic-educational course. *International Archives of Occupational and Environmental Health* 71(5): 336–342. doi:10.1007/s004200050289.

Frohm, J. 2008. Levels of automation in production systems. Ph.D. dissertation, Chalmers University of Technology.

Goggins, R.W., P. Spielholz, and G.L. Nothstein. 2008. Estimating the effectiveness of ergonomics interventions through case studies: Implications for predictive cost-benefit analysis. *Journal of Safety Research* 39(3): 339–344. doi:10.1016/j.jsr.2007.12.006.

Krüger, J., T.K. Lien, and A. Verl. 2009. Cooperation of human and machines in assembly lines. *CIRP Annals-Manufacturing Technology* 58(2): 628–646. doi:10.1016/j.cirp.2009.09.009.

Krüger, J., and T.D. Nguyen. 2015. Automated vision-based live ergonomics analysis in assembly operations. *CIRP Annals—Manufacturing Technology, May.* doi:10.1016/j.cirp.2015.04.046.

Lämkull, D., L. Hanson, and R. Örtengren. 2009. A comparative study of digital human modelling simulation results and their outcomes in reality: A case study within manual assembly of automobiles. *International Journal of Industrial Ergonomics* 39(2): 428–441. doi:10.1016/j. ergon.2008.10.005.

Nguyen, T.D., C. Bloch, and J. Krüger. 2016. The working posture controller: Automated adaptation of the work piece pose to enable a natural working posture. *Procedia CIRP* 44: 14–19. doi:10.1016/j.procir.2016.02.172.

Nguyen, T.D., M. Kleinsorge, and J. Krüger. 2014. ErgoAssist: An assistance system to maintain ergonomic guidelines at workplaces. Paper presented at the IEEE Conference on Emerging Technology and Factory Automation (ETFA), Barcelona, September 16–19.

Paul, P., P.P.F. Kuijer, B. Visser, and H.C. Kemper. 1999. Job rotation as a factor in reducing physical workload at a refuse collecting department. *Ergonomics* 42(9): 1167–1178. doi:10. 1080/001401399185054.

Schaub, K., G. Caragnano, B. Britzke, and R. Bruder. 2013. The European Assembly worksheet. *Theoretical Issues in Ergonomics Science* 14(6): 616–639. doi:10.1080/1463922X.2012. 678283.

Schmidtler, J., C. Hölzel, V. Knott, and K. Bengler. 2014. Human centered assistance applications for production. In *Advances in the ergonomics in manufacturing: Managing the enterprise of the future,* 13 ed. S. Trzcielinski, and W. Karwowski, 380–391. Krakow: AHFE Conference.

Schneider, E., X. Irastorza, and S. Copsey. 2010. *OSH in figures: Work-related musculoskeletal disorders in the EU—Facts and Figures.* Luxembourg: Office for Official Publiction of the European Communities.

United Nations—Department of Economic and Social Affairs. 2007. *Indicators of sustainable development: Guidelines and methodologies.* New York: United Nations Publications.

van der Molen, H.F., J.K. Sluiter, C.T.J. Hulshof, P. Vink, and M.H.W. Frings-Dresen. 2005. Effectiveness of measures and implementation strategies in reducing physical work demands due to manual handling at work. *Scandinavian Journal of Work, Environment & Health* 10 (suppl 2): 75–87. PMID:16363450.

Vignais, N., M. Miezal, G. Bleser, K. Mura, D. Gorecky, and F. Marin. 2013. Innovative system for real-time ergonomic feedback in industrial manufacturing. *Applied Ergonomics* 44(4): 566–574. doi:10.1016/j.apergo.2012.11.008.

Weidner, R., N. Kong, and J.P. Wulfsberg. 2013. Human hybrid robot: A new concept for supporting manual assembly tasks. *Production Engineering* 7(6): 675–684. doi:10.1007/ s11740-013-0487-x.

Part III
Solutions for Sustainable Product Development

T. Buchert, Chair of Industrial Information Technology, Institute for Machine-tools and Factory Management, Technische Universität Berlin, Berlin, Germany

As shown in the previous book part, innovations in manufacturing tech-nologies have the potential for significant reduction of resource consumption as well as for decreasing health related workplace-risks at the same time. Nevertheless, once a product is manufactured its sustainability performance along the whole lifecycle is already determined to a large degree. In this context the product design stage can be seen as a major lever which defines for example necessary manufacturing steps, longevity of product usage and potentials for material recovery once the product is disposed.

A sustainable design starts by limiting potential harmful effects of the product along its whole lifecycle for various stakeholders. Classical exam-ples in this context are gaseous emissions contributing to anthropogenic climate change, toxic liquid and solid waste or unnecessary cost for the company, customers or the society. Despite the prevention of negative effects sustainable products also provide opportunities to fulfil human needs and provide value in all areas of human living (in particular mobil-ity, production and energy). Solutions in this context comprise for exam-ple sophisticated highly quality products making everyday life easier, frugal innovations to address basic needs in developing countries or mechanisms for fostering societal cohesion by including people with disabilities or the elderly.

Sustainable Product Development characterizes the science and art of foreseeing the whole product lifecycle by handling multiple decision criteria at the same time to find a compromise between all involved stakeholders including the company, society, environment and future generations. Hence, research on sustainable pro-duct development focuses on a diverse set of research questions of which some are listed below in an exemplary manner:

I.What constitutes a sustainable product?

II.How can sustainability be integrated into the design/design management process?

III.Which forms of decision support are necessary to enable stake-holders for sustainable product development?

Research on these questions under the label of sustainability is conducted since approximately 10 years making it a relatively new area of research. However, since sustainable product development is grounded in the field of Ecodesign there are almost 30 years of experience comprising a massive amount of publications, industrial application cases and a large variety of tools and methods which have been developed in that context.

The first contribution to this book part will take a closer look at how the research field evolved with the years from pure Ecodesign to an inte-grated view of sustainable product and business model design enabling the transition to a circular economy. Furthermore, an outlook is given how the journey will continue in the future and what will be the main challenges to solve before sustainable value creation can be achieved from a product development perspective.

An example how the three above-mentioned questions can be addressed in a corporate context can be found in the second contribution of this book part. Here it is discussed how the rather fuzzy concept of sustainability can be integrated into conventional product development processes in producing companies. In this context the target-driven approach for Sustainable Product Development searches for ways to increase transparency of decision making. After naming the challenges for definition and validation of sustainability targets options for decision support are presented. The approach utilizes software support which is embedded into existing engineering IT tools.

The third contribution focuses on the end of life phase of the product lifecycle which recently gained increased attention through research on enabling a circular economy. The main challenge in this context is to guarantee that added-value, embedded in a manufactured product can be conserved after its first utilization period. Additional usage phases can be achieved through direct reuse, remanufacturing or repurposing. Since these end of life options need to be considered in product design already possible options for implementation are discussed and practical guidelines are presented.

From Ecodesign to Sustainable Product/Service-Systems: A Journey Through Research Contributions over Recent Decades

Tim C. McAloone and Daniela C.A. Pigosso

Abstract Corporate approaches towards sustainability integration into product development have significantly evolved since the early 1990s. Ecodesign, defined as the integration of environmental issues into product development, arose in the 1990s as a key concept for the enhancement of products' environmental performance. An intense development of ecodesign methods and tools could be observed in the 1990–2010 period, leading to successful pilot cases in industry, in which environmental gains were demonstrated. In the 2010s, the need for a systems perspective to solve the environmental crisis has been highlighted, and the concept of product/service-systems started to gain momentum due to the high potential for enhanced environmental performance and improved competitiveness, by means of new business models and dematerialization. Recently, a transition towards Circular Economy and the integration of social innovation into sustainability initiatives can be observed, which leads to strategic and holistic sustainability considerations in the design of complex systems. In this chapter, the evolution of sustainability concepts and their integration into product development is presented and exemplified in three periods: 1990–2010; 2010–2020 and 2020–2030. While the first two periods present the actual development of the field, the last period represents the evaluation and projection of the trends developed by the authors. By analysing the three periods, the authors aim to discuss the journey from ecodesign to sustainable product/service-systems over the last decades, experienced by academia and practitioners, and to highlight their views on how the field is going to develop over the next 10 years.

Keywords Ecodesign · Product/service-systems · Sustainable innovation · Circular Economy

T.C. McAloone (✉) · D.C.A. Pigosso
Technical University of Denmark, Kongens Lyngby, Denmark
e-mail: tmca@dtu.dk

© The Author(s) 2017
R. Stark et al. (eds.), *Sustainable Manufacturing*, Sustainable Production,
Life Cycle Engineering and Management, DOI 10.1007/978-3-319-48514-0_7

99

1 Introduction

Since the early 1990s academics and practitioners have been placing increasing focus on sustainability awareness in the product development process, by means of tools, methods and targeted projects towards sustainability enhancement. In the early years, the focus was on designing better for environmental concerns, from which period we see the beginnings of what today is a huge catalogue of approaches towards life cycle assessment and ecodesign, to name just two of the very popular environmental improvement approaches. The important questions are: how does it look today? In which dimensions have we developed our knowledge? How has the world changed since we began to work with ecodesign? And are we effectively developing our competencies, in order to be more effective in our approach to continued sustainability enhancement?

In an attempt to answer the above questions, the authors have carried out a review and reflection of the previous and current decades, before projecting our thoughts onto what we see may be the foreseeable future for sustainability enhancement through business- and product development. To help to make this reflection, three time periods and nine dimensions have been identified, so as to characterize the general sustainability focus, over time. The time periods in focus are 1990–2010 (characterised as the rise and establishment of ecodesign); 2010–2020 (a systems perspective on ecodesign); and 2020–2030 (perspectives for a sustainable and Circular Economy). The nine dimensions identified for the review and reflection exercise were the following:

- **Main goals/objectives:** This dimension was included to highlight what was the main sustainability design object of the company, in the given time period, ranging from very concrete artefact-focused objectives to more cognitive objectives seen in more recent times.
- **Expected results:** This denotes the main focus of industry/society in each given time period, also indicating the level of proactivity towards sustainability within the given period.
- **Main aim:** This dimension marks whether the main aim of the sustainability effort is towards building, implementing, or fully integrating tools into the organisation.
- **Basic approach:** This dimension helped the authors to differentiate, whether the general approach to sustainability improvement could be characterised as being singular problem-focused, system-oriented, or holistic.
- **Envisaged cost-benefit:** The general attitude of industry, towards sustainability's value contribution was charted in this dimension, to provide a candid image of the general level of expectation towards sustainability.
- **Sustainability ambition:** This dimension denotes which combination of the three so-called pillars of sustainability (environmental-social-business) were most in focus in the given time period.

- **Business mindset:** This dimension was included in order to differentiate between incumbent take-make-waste (or 'linear') business mindsets, or whether a more circular mindset was evident in a given time period.
- **What are we changing:** This dimension was added to place focus on what the main objective of sustainability efforts typically was in a given time period, whether it be to make direct product improvements, more systemic process improvements, or a generally holistic focus on the competencies of the professionals in the product development organisation.
- **Decision-making level:** This final dimension was used to mark which dominant part of the organisation was most instrumentally being engaged, in a given time period.

The following sections review and reflect on the activities, campaigns, research, industry examples, and key results gained from each of the three respective time periods. The above sustainability dimensions are used as way of structuring this reflection. A progression and a development can be observed, in the three time periods considered.

2 1990–2010: The Rise and Establishment of Ecodesign

Over the 1990–2010 period, companies have significantly evolved their approaches towards the integration of sustainability into their business activities, developing from a passive and reactive stance, towards the adoption of more preventive and proactive approaches.

The business concern related to sustainability issues in this period was directly related to the intensification of environmental awareness in the 1970s and 1980s. The increased awareness was a consequence of the pollution caused by a generally passive attitude until then adopted by industry, where almost no mechanisms for pollution control were in place.

Within the passive approach, industrial waste generated in the production processes by manufacturing companies was disposed directly in the environment without any kind of treatment, leading to a severe pollution of the environmental compartments (soil, air and water) and causing serious damage to both human health and quality of life.

In recognition of the pollution effects on human health and the environment, governments worldwide started to intensify their environmental legislation programmes in the 1980s, which aimed at regulating companies' activities concerning pollution control. From this development and strengthening of environmental legislation, companies started to shift from a passive stance towards the adoption of what we today would call reactive approaches, which focused on the so-called 'end-of-pipe' solutions.

The 'end-of-pipe' solutions aimed at reducing the pollution potential of industrial waste, so as to comply with the enacted legislation, by investing in technologies, which were chiefly intended for the treatment of industrial wastewater, solid waste and gases generated in the production processes. Due to the relatively high investments for the implementation of 'end-of-pipe' solutions, there was a strong tendency to understand environmental and sustainability issues as a cost to the organization, rather than as an opportunity.

In the early 1990s, a preventive approach emerged in a context in which companies started to improve their manufacturing processes, in order to minimize the increasing costs related to 'end-of-pipe' solutions, to comply with the ever-constraining legislation and to increase resource efficiency. Concepts such as Pollution Prevention and Cleaner Production were key in the period, when the preventive approach was at its highest. The aim was to reduce the waste generation directly at its source, i.e. in the production processes, thereby reducing treatment and final disposal costs (UNEP 2004; Ahmed 2012).

Besides being driven by legal aspects, this change in attitude was also due to the recognition of the real costs associated to the traditional 'end-of-pipe' approaches. In addition to the costs usually attributed to treatment and disposal, there are other costs that are usually not taken into account, such as, for example, costs related to the loss of resources (raw materials, water, energy, etc.), legal and regulatory non-compliance, corporate image, to name a few. Typically, for every dollar accounted for waste treatment or disposal, a further two to three dollars are 'hidden' or simply ignored, even in well managed and large companies (UNEP 2004).

Despite the innumerous benefits of reactive and preventive approaches to sustainability enhancement, they alone are not enough to deal with the sustainability challenges that our society was—and still is—facing, due to the ever-increasing production and consumption of products.

In the late 1990s, the recognition that products were at the origin of most of the pollution and resource depletion caused by our society became evident and a transition to a more proactive approach could be observed. At that time, companies started to realize that all products caused some sort of impact, not only during the manufacturing processes, but also throughout their entire life cycles, from raw material extraction through manufacturing, use and final disposal (Fava 1998).

In this context, ecodesign emerged as a promising approach for the integration of environmental considerations in product development processes, where the opportunities for enhancement of the environmental performance across the product life cycle was estimated to be around 80 % (through the definition of materials, suppliers, product performance, etc.) (Mcaloone and Bey 2009). The introduction of the life cycle thinking was associated with efforts to increase efficiency throughout the product life cycle (Brezet et al. 1999; Sherwin and Bhamra 1999; Stevels et al. 1999).

To enable ecodesign implementation in companies, several methods and tools were developed by industry and academia in this period. Several approaches for the

Fig. 1 The rise and establishment of ecodesign (1990–2010)

evaluation of the environmental performance of products (e.g. through Life Cycle Assessment (LCA) and similar approaches) were developed and ecodesign guidelines for enhanced environmental performance of products were consolidated for different product types and industrial sectors (Caspersen and Sørensen 1998; Brezet et al. 1999).

The basic approach at this moment was focused on specific product issues (e.g. minimization of weight, elimination of hazardous substances, enhancement of energy efficiency, etc.). At this time, and due to the previous experience with end-of-pipe approaches, which were costly and mainly there for legislative compliance, sustainability was chiefly viewed as a necessary cost, with only very few companies being able to demonstrate the business benefits linked with ecodesign implementation.

The take-make-waste paradigm of the linear economy was the main paradigm in most of the companies at this time, although initial discussions regarding the impacts and importance of the end-of-life of products started to enhance towards the end of the 2010s (Rose et al. 2002). Most of the actions taken for ecodesign implementation were at an operational level, looking mainly at the product level and from a strict design perspective, linked to material and energy efficiency.

By the end of the 2010s, more than 100 different methods and tools were developed, but the broad uptake by industry was not as expected (Baumann et al. 2002) and new challenges started to be observed by society (Pigosso et al. 2015). At that time, there was a need to evolve the ecodesign concepts and allow for a broader implementation and uptake by industry. Figure 1 provides a summary of the main characteristics of corporate sustainability in the 1990–2010 period.

3   2010–2020: A Systems Perspective on Ecodesign

In the period 2010–2020 (which encapsulates the current time of writing), a shift can be observed in society, away from the more reactive, tool-building and singular problem-focus of the first era. In this period, a new wave of globalization is in full flow, enabled by technology and near-instant availability of products and services, all around the world. As the world gets smaller, so to speak, singular products often become commoditized, with their perceived value reducing to a minimum. For instance, the increased rate of commoditization can seem like a vastly negative trend, environmentally, due to ever-shortening product lifetimes and large bouts of waste, within a linear economy. However, two counter-developments have emerged, namely the embedding of high value in high quality products; and the emergence of product/service-systems onto the market. High-value, high-quality products (e.g. premium-priced smartphones and high-end portable computers) indeed provide some of the answer to the previous era's problem with commoditization and product waste. Product/service-systems, PSS (which effectively are purposely co-developed product and service bundles) are also increasingly normal in both B2B and B2C markets. PSS come with new business models, which often focus on providing more value-add from one installed base of a product, by means of some form of product life extension (often through sharing), and therefore dematerialization of the physical artefacts, which are component parts of the PSS under offer (Bey and Mcaloone 2006).

It is in this time period that many companies are starting to formulate sustainability goals, together with ways in which these will be measured, be they environmental, social and/or business-oriented. The very intensive period of tool building has slowed in this decade, with more emphasis being placed on how to actually successfully select from the large lists of tools and methods and implement the most suitable tools within the company (Pigosso et al. 2011; Bovea and Perez-Belis 2012). This is a positive development, as we can identify over 800 ecodesign best practices already (Pigosso et al. 2014)—the focus must now be on how to ensure successful implementation of these tools and methods into the business- and product development processes of the enterprise.

Together with the shift from products to PSS as a standard sustainability design object, the basic approach has shifted, so as to incorporate more sustainable decision points at a given time, thereby encompassing a systems approach towards sustainability enhancement. Nevertheless, many companies are not yet realizing the full benefit of their efforts towards sustainability improvement, often rendering sustainability as an activity that may not any longer be seen as a net cost to the company, but is still not a sufficient value-creator in itself.

In this decade, social sustainability is a clear focus point for the organization and a number of projects (often in collaborations between academia and enterprises) have been completed, where social sustainability methods and metrics have been developed, tried and tested (Ny et al. 2006; Boström 2012).

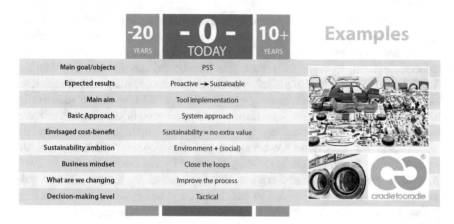

Fig. 2 A systems perspective on ecodesign (2010–2020)

Western society is beginning to pay increasing attention in this decade to closing loops, rather than operating in a linear economy. Focus is increasingly being placed on takeback schemes, Design for Recycling activities, new business models to revalorize waste, and new forms of artefact sharing systems (e.g. bike-sharing, car-sharing, tool-sharing, to name but a few) (McDonough and Braungart 2010). We are by no means circular in our approach, but closed loop activities are beginning to be favoured over linear economy activities.

Looking inside companies and universities, we can see increasing focus being placed on how to create better processes towards sustainable product development, rather than simply creating yet another tool or a method. With this elevation of activities to the level of PSS, systems thinking and closed loop operations, companies are increasingly engaging the middle-management (tactical) levels of their business- and product development activities, in order to understand how to leverage greater parts of the companies' value-adding activities, through more tactical deployment of sustainability thinking (Tukker 2004). Figure 2 shows a summary of the main characteristics of corporate sustainability in the 2010–2020 period.

4 2020–2030: Perspectives for a Sustainable and Circular Economy

An even more significant transition to corporate sustainability is expected in the upcoming decade. Although predicting the future is impossible, we have attempted to develop a scenario of how current initiatives might possibly deploy over the next decade, based on an analysis of current trends and past developments.

Increasing recognition of the need to mitigate the effects of population growth, wealth increase and human consumption is currently leading several international organizations to consensually highlight the need for a significant change in our economic system, in order to respect planetary boundaries (Steffen and Stafford Smith 2013; Häyhä et al. 2016). Some examples of sustainability-related initiatives include: the roadmap for developing energy efficient and low-carbon societies by 2050, developed by the European Union; the 'green growth' framework to foster economic growth while ensuring the availability of natural resources, by the Organisation for Economic Co-operation and Development (OECD); and the Sustainable Development Goals (SDGs), launched by the United Nations in 2016. In order to reach global and European development goals, the private and governmental sectors in Europe need to undergo a large and systemic transition.

Due to the recognition of the systemic sustainability challenge faced by our society, a change towards extended collaboration within and across value chains is expected. Collaboration must be focused on developing new solutions and economic systems, bringing together different stakeholders in society, that help addressing the planetary boundaries (Steffen and Stafford Smith 2013).

An increasing amount of businesses will be maturing their approaches towards sustainability and increasingly integrating sustainability into not just the high-level strategic goals of the company, but also the everyday business and product development processes. This will allow each and every decision in the organization to be taken based on solid and conscious sustainability considerations. It will also give rise to a holistic approach, in which the connections and interfaces among complex systems are considered and their dynamic natures understood.

Competences will be significantly enhanced to be able to cope with the understanding of complex problems and the collaborative development of efficient solutions. Sustainability will be defined and committed at a strategic level in organizations and the deployment into the tactical and operational levels will be enabled by the enhanced maturity of companies on sustainability enhancement.

At this point, companies will have the contents and the context to be able to understand that sustainability equals business, and that there is no other alternative way of being successful in a business context. In fact, such signs are already evident in the very leading-edge corporations, which have put a direct relationship between sustainability and business-enhancing innovation (Ellen MacArthur Foundation 2015a). First on achieving a critical mass of this type of company, recognizing the opportunities of business-driven sustainability action, will we see that the sustainability concept defined as the balance between the environmental, social and economic dimensions will finally be fully met.

In the next decade, problems and risks related to resource scarcity and product disposal will be minimized by an enhanced uptake of the concept of Circular Economy (Ellen Macarthur Foundation et al. 2015), which is currently being boosted in many parts of the world.

Circular Economy is increasingly seen as a key approach to operationalizing goals and supporting the transition by enhancing competitiveness, economic growth and sustainability in many parts of modern society. Circular Economy is defined by the Ellen MacArthur Foundation as *"an economy that provides multiple value creation mechanisms, which are decoupled from the consumption of finite resources"* (Ellen MacArthur Foundation 2015a). Unlike the traditional linear 'take-make-waste' approach, the goal of Circular Economy is to seek to respect planetary boundaries through increasing the share of renewable or recyclable resources, whilst reducing the consumption of raw materials and energy and thus bringing down emissions and material losses (EEA 2016). Creating a Circular Economy requires fundamental changes throughout the value chain, from innovation, product design and production processes all the way to end of life, new business models and consumption patterns (EEA 2016).

Large and established, as well as small and start-up players in the industry are increasingly recognizing the need to commercialize secondary raw materials, to ensure spare-parts availability and to actively begin to devise alternative and innovative business models, disruptive to their current ways of working (2016). Among the strategies being addressed are: expansion of high value-added services; focus on Total Cost of Ownership (TCO) over the product lifetime; outsourcing agreements and rental offerings; technical leadership; and optimized product quality. Manufacturers are increasingly positioning their offerings, such as equipment financing; training for the best use of machines; fleet management; and equipment relocation services, as ways in which to enhance their value propositions to their customers. The positive news is that these new value propositions by the industry are potential components of a circular business model approach.

A successful transition to Circular Economy requires a systemic change in the way companies understand and do business, with sustainability as a strong foundation. Circular Economy will be enabled by the combined application of three component elements: (i) Business Model Innovation; (ii) Sustainable Design and Ecodesign; and (iii) Internet of Things coupled with Digital Transformation.

One of the most powerful enablers of a circular economy is sustainable business model innovation (Chun and Lee 2013; Pigosso and McAloone 2015; Reim et al. 2015). Business models that successfully incorporate Circular Economy principles have a direct and lasting effect on the social, economic and environmental systems (EEA 2016). Taking a sustainable business model view on Circular Economy promotes the integration of suitable approaches such as ecodesign, reuse, sharing, leasing, repair, refurbishment and recycling. By integrating the most suitable of these approaches to one's business- and product development will play a significant role in maintaining the utility of products, components and in realizing circular business models (EEA 2016).

Circular Economy business models can only be realized by the development of products, services and Product/Service-Systems that can be easily disassembled, remanufactured, recycled and reused (Bakker et al. 2014; Tukker 2015). Common approaches for the design of circular products includes the application of Design for Recycling, Design for Remanufacturing and Design for Disassembly methods, tools and guidelines (Sundin and Bras 2005; Pigosso et al. 2010; Achillas et al. 2013). Nevertheless, in order to ensure a superior sustainability performance of products, the entire life cycle of products need to be considered.

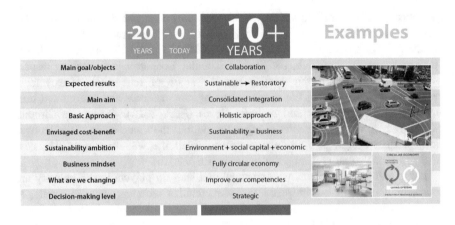

Fig. 3 Perspectives for a sustainable and circular economy (2020–2030)

Circular Economy can benefit greatly by equipping products with intelligence, so that they can adapt and respond to change and remain fit-for-purpose over longer time periods (Ellen MacArthur Foundation 2015b). A whole new range of virtual services and sharing economy platforms support the prolonged technical lifetime (and sometimes also up-cycling) of products by monitoring the condition of individual components or whole product systems.

In this context, Circular Economy will lead to the development of innovative business models, products, value chains, partnerships, and technologies that will enable a much more and efficient closed loop of materials and energy—and ultimately a more robust economy.

Due to the significant undermining of planetary boundaries caused through the industrial activities of the past century, it is increasingly recognized that the sustainability concept will need to embrace restoratory concepts, so as to reestablish the planetary boundaries at safe levels and not undermine life on Earth (Fig. 3).

5   Summary and Final Remarks

This chapter has provided our reflection of the development and evolution of sustainability initiatives and approaches observed since the 1970s in a corporate context. The reflection has structured in three distinct periods, which are characterized by their own specificities, challenges and focus areas (Fig. 4).

Despite the common perception that we are still struggling with the same issues since the early stages of corporate sustainability initiatives, a clear change in patterns and a significant evolution of the discussion is observed. Governmental bodies, universities, non-governmental organizations, companies and the civil society have significantly raised and enriched the debate around sustainability.

	-20 YEARS	- 0 - TODAY	10+ YEARS
Main goal/objects	Product	PSS	Collaboration
Expected results	End-of-pipe → proactive	Proactive → Sustainable	Sustainable → Restoratory
Main aim	Tool building	Tool implementation	Consolidated integration
Basic Approach	Singular problem approach	System approach	Holistic approach
Envisaged cost-benefit	Sustainability = cost	Sustainability = no extra value	Sustainability = business
Sustainability ambition	Environment	Environment + (social)	Environment + social capital + economic
Business mindset	Linear economy	Closing the loops	Fully circular economy
What are we changing	Improve the product	Improve the process	Improve our competencies
Decision-making level	Operational	Tactical	Strategic

Fig. 4 A journey through research contributions over the recent decades (1990–2030)

Furthermore, industry interest and uptake at the strategic, tactical and operational levels is following a steady increase—although many challenges are still faced for full sustainability integration.

In order to be able to cope with the sustainability challenges faced by our society and respecting the planetary boundaries, the speed of change and actual uptake by industry and a varied set of stakeholders must enhance significantly over the next decade. At the same time that ambitious targets must be set, it is important that industry companies take a systematic and step-by-step approach towards enhancing their organizational maturity to be able to develop and perpetuate successful and sustainable businesses.

References

Achillas, C., D. Aidonis, C. Vlachokostas, et al. 2013. Depth of manual dismantling analysis: A cost-benefit approach. *Waste Management* 33: 948–956. doi:10.1016/j.wasman.2012.12.024.

Ahmed, K. 2012. *Getting to green : A sourcebook of pollution management policy tools for growth and competitiveness*. Washington, DC.

Bakker, C., F. Wang, J. Huisman, and M. den Hollander. 2014. Products that go round: Exploring product life extension through design. *Journal of Cleaner Production* 69: 10–16. doi:10.1016/j.jclepro.2014.01.028.

Baumann, H., F. Boons, and A. Bragd. 2002. Mapping the green product development field: Engineering, policy and business perspectives. *Journal of Cleaner Production* 10: 409–425. doi:10.1016/S0959-6526(02)00015-X.

Bey, N., and T.C. Mcaloone. 2006. From LCA to PSS—making leaps towards sustainability by applying product/ service-system thinking in product development. In *13th CIRP International Conference on Life Cycle Engineering*, 571–576.

Boström, M. 2012. A missing pillar? Challenges in theorizing and practicing social sustainability: Introduction to the special issue. *Sustainability: Science, Practice, Policy* 8: 1–13.

Bovea, M.D., and V. Perez-Belis. 2012. A taxonomy of ecodesign tools for integrating environmental requirements into the product design process. *Journal of Cleaner Production* 20: 61–71. doi:10.1016/j.jclepro.2011.07.012.

Brezet, H., A. Stevels, and J. Rombouts. 1999. LCA for EcoDesign : The Dutch Experience Ab Stevels design for sustainability program Delft University of Technology. In *First International Symposium On Environmentally Conscious Design and Inverse Manufacturing, 1999. Proceedings. EcoDesign'99*, 36–40.

Caspersen, N.I., and A. Sørensen. 1998. Improvements of products by means of lifecycle assessment: High pressure cleaners. *Journal of Cleaner Production* 6: 371–380.

Chun, Y.-Y., and K.-M. Lee. 2013. Life cycle-based generic business strategies for sustainable business models. *Journal of Sustainable Development* 6: 1–15. doi:10.5539/jsd.v6n8p1.

EEA. 2016. *Circular economy in Europe: Developing the knowledge base*. Luxembourg.

Ellen MacArthur Foundation. 2015a. *Growth within: A circular economy vision for a competitive Europe*, 100.

Ellen MacArthur Foundation. 2015b. *Intelligent assets: Unlocking the circular economy potential*.

Ellen Macarthur Foundation, SUN, Environment MC for B. 2015. *Growth within: A circular economy vision for competitive Europe*.

Fava, J.A. 1998. Life cycle perspectives to achieve business benefits: From concept to technique. *Human and Ecological Risk Assessment* 4: 1003–1017. doi:10.1080/10807039891284947.

Häyhä, T., P.L. Lucas, D.P. van Vuuren, et al. 2016. From Planetary Boundaries to national fair shares of the global safe operating space—How can the scales be bridged? *Global Environmental Change* 40: 60–72. doi:10.1016/j.gloenvcha.2016.06.008.

Mcaloone, T.C., and N. Bey. 2009. *Environmental improvement through product development—A guide*, 1st ed. Copenhagen: Denmark.

McDonough, W., and M. Braungart. 2010 *Cradle to cradle: Remaking the way we make things*. Macmillan.

Ny, H., J.P. MacDonald, G. Broman, et al. 2006. Sustainability constraints as system boundaries—An approach to making life-cycle management strategic. *Journal of Industrial Ecology* 10: 61–77. doi:10.1162/108819806775545349.

Pigosso, D.C., and T.C. McAloone. 2015. Supporting the development of environmentally sustainable PSS by means of the ecodesign maturity model. *Procedia CIRP* 30: 173–178. doi:10.1016/j.procir.2015.02.091.

Pigosso, D.C.A., T.C. McAloone, and H. Rozenfeld. 2014. Systematization of best practices for ecodesign implementation. In *International Design Conference—DESIGN 2014*, 1651–1662. Croatia: Dubrovnik.

Pigosso, D.C.A., T.C. McAloone, and H. Rozenfeld. 2015. Characterization of the state-of-the-art and identification of main trends for Ecodesign Tools and Methods: Classifying three decades of research and implementation. *Journal of the Indian Institute of Science* 95: 405–427.

Pigosso, D.C.A., H. Rozenfeld, and G. Seliger. 2011. Ecodesign Maturity Model : Criteria for methods and tools classification. *Advanced Sustainable Manufacturing* 241–245.

Pigosso, D.C.A., E.T. Zanette, A.G. Filho, et al. 2010. Ecodesign methods focused on remanufacturing. *Journal of Cleaner Production* 18: 21–31. doi:10.1016/j.jclepro.2009.09.005.

Reim, W., V. Parida, and D. Örtqvist. 2015. Product-service systems (PSS) business models and tactics—a systematic literature review. *Journal of Cleaner Production* 97: 61–75. doi:10.1016/j.jclepro.2014.07.003.

Rose, C.M., K. Ishii, and A. Stevels. 2002. Influencing design to improve product end-of-life stage. *Concurrent Engineering: Research and Applications* 13: 83–93. doi:10.1007/S001630100006.

Sherwin, C., and T. Bhamra. 1999. Beyond engineering: ecodesign as a proactive approach to product innovation. In *Proceedings of the First International Symposium on Environmentally Conscious Design and Inverse Manufacturing*, 41–46. doi:10.1109/ECODIM.1999.747578.

Steffen, W., and M. Stafford Smith. 2013. Planetary boundaries, equity and global sustainability: Why wealthy countries could benefit from more equity. *Current Opinion in Environmental Sustainability* 5: 403–408. doi:10.1016/j.cosust.2013.04.007.

Stevels, P.A., B. Sk, P.O. Box, and N.-J.B. Eindhoven. 1999. *Integration of EcoDesign into business, a new challenge*, 27–32.

Sundin, E., and B. Bras. 2005. Making functional sales environmentally and economically beneficial through product remanufacturing. *Production* 13: 913–925. doi:10.1016/j.jclepro.2004.04.006.

Tukker, A. 2004. Eight types of product–service system: Eight ways to sustainability? Experiences from SusProNet. *Business Strategy and the Environment* 13: 246–260. doi:10.1002/bse.414.

Tukker, A. 2015. Product services for a resource-efficient and circular economy—A review. *Journal of Cleaner Production* 97: 76–91. doi:10.1016/j.jclepro.2013.11.049.

UNEP. 2004. Guidelines for the integration of cleaner production and energy efficiency.

(2016) European Forum for manufacturing debates circular economy. http://www.cece.eu/news-and-events/news-details/article/european-forum-for-manufacturing-debates-circular-economy/.

Design for High Added-Value End-of-Life Strategies

Tom Bauer, Daniel Brissaud and Peggy Zwolinski

Abstract Sustainable manufacturing is a rising issue. Ensuring both consumer satisfaction and minimal environmental impact is very challenging. In that whole process, it is customary to say that the design stage determines 80 % of the future environmental impact. One way to contain this impact at an acceptable level is to manage the products' end-of-life from the design activities. This chapter points out product reuse strategies—i.e. *direct reuse and remanufacturing*—aiming at conserving the added-value of used products as much as possible into new products. The first contribution attempts to provide a state-of-the-art of design for these high added-value end-of-life strategies. Direct reuse and remanufacturing are thus analysed and the principal design guidelines are furthermore given, classified according to three dimensions: product, process and business model. This chapter then contributes to enlarging the spectrum of reuse strategies, presenting an innovative end-of-life strategy: repurposing. It consists of reusing products in other applications after transformations. The main challenges of such a strategy will be discussed.

Keywords Design for X · End-of-life strategy

1 Introduction

There is a need to improve the environmental orientation of products and the management of their end-of-life (EoL) represents one way of achieving this. Many studies argue that it could be initiated from different actors: customers, pushing for greener products; companies, willing to reduce the environmental footprint of their products as much as increasing their revenues; or regulation, favouring

T. Bauer (✉) · D. Brissaud · P. Zwolinski
Univ. Grenoble Alpes, CNRS, G-SCOP, 38000 Grenoble, France
e-mail: tom.bauer@grenoble-inp.fr

© The Author(s) 2017
R. Stark et al. (eds.), *Sustainable Manufacturing*, Sustainable Production,
Life Cycle Engineering and Management, DOI 10.1007/978-3-319-48514-0_8

low-impact-products and obligating producers to handle their end of life processes, beginning with the design phase (Global Reporting Initiative 2013; Goodall et al. 2014).

An end-of-life strategy refers to the manner in which one manages the product right after its user has discarded it. The focus today is on end-of-life strategies that maximise the value of the products, so-called reuse strategies. These strategies have key characteristics that must guide designers to facilitate their initial setup. This chapter tries first of all to make these strategies clear as well as outline what the drivers are for the most adapted designs. An exploration follows, of how the main end-of-life strategies maximise the value of products, along with how to support product designers in their willingness to pursue these maximizing-value strategies.

These end-of-life strategies and their consequences on the design of products are now well-known and shared among companies: the product characteristics, its performances and the recovering process are described in literature. Nevertheless, the discussion is open to proposing new strategies that retain more and more added value of used products for the purpose of ultimately manufacturing innovative products. Repurposing, meaning that end-of-life products can be revamped into different applications than the former ones to prolong their lifetime, needs now to be understood, modelled and analysed in pursuit of guaranteeing its implementation and its potential value.

Following this introduction, the chapter describes the product end-of-life strategies in Sect. 2, before focusing on high added-value strategies, and reuse strategies, in Sect. 3. They will be described in terms of product, process and business model characteristics and an overview of the main guidelines for assisting the product design work will be summarized. Section 4 paves the way for the repurposing strategy to be presented and discussed.

2 High Added-Value End-of-Life Strategies

The need to define a product end-of-life strategy takes place when the product is considered as a 'waste' (European Commission 2008). The European Commission (2008) defines *waste* as: "any substance or object which the holder discards or intends or is required to discard." Depending on its type, characteristics and working conditions, the discarded product may follow one or another strategy. ISO proposes a classification of end-of-life strategies though the standard 14062 (ISO 2002), which has been ranked depending on potential environmental gains: (a) prevention, (b) reuse, (c) recycling, (d) energy recovery and (e) disposal; (European Commission 2008). In this chapter, the focus is set on strategies which aim at maintaining as much added-value in products as possible.

First of all, energy recovery and landfilling do not represent sustainable strategies since they do not recover any element of the products: both added value and material are destroyed. These strategies will be grouped under the "waste" label in the chapter (see 1 in Fig. 1). Recycling (see 2 on Fig. 1) consists of recovering

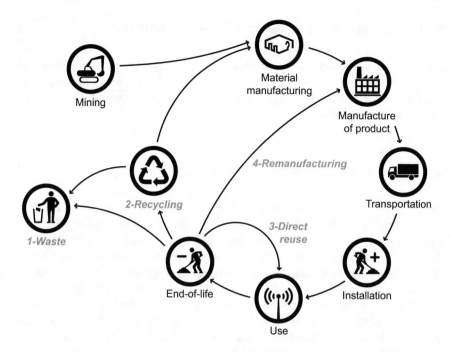

Fig. 1 Product lifecycle and the 4 main end-of-life strategies (adapted from Zhang 2014)

materials from the discarded products in order to avoid new raw material extraction and, in so doing, limit the environmental impact and supply issues. The recycling strategy destroys the added-value of the product and instead only recovers materials. The strategies that recover material and retain the product's added-value are called reuse strategies. It can be split in two distinct sub-strategies: direct reuse and remanufacturing. Direct reuse (see 3 on Fig. 1) is a process where the quality of the product and the market conditions allow for continued use of the same product by another customer. The remanufacturing strategy (see 4 on Fig. 1) concerns products that have to go through a new manufacturing process before being put back on the market. Indeed, direct reuse and remanufacturing both aimed at providing as-new products with at least the same guaranties and performances as a new product and for the same application. Finally, prevention mainly consists of avoiding the impact before the end of the product life, by minimizing wastes.

The paper focuses on end-of-life strategies that conserve added-value of products, meaning the materials after manufacturing transformation. These strategies are called "reuse strategies." The "quantity" of added-value retained, and the corollary "quantity" of transformation needed to recover the added-value missing, characterize the process of remanufacturing of the product from "high added-value retained—light remanufacturing process" (direct reuse strategy) to "less but real added-value kept—standard remanufacturing process" (remanufacturing strategy).

3   Design for Direct Reuse and Remanufacturing

The focus of this section is on reuse strategies happening right after the End-of-Use (EoU) of products. A distinction is made between Design for direct Reuse (DfdR) and Design for Remanufacturing (DfRem). Definitions, explanation and design guidelines are pointed out.

3.1   Definitions and Main Characteristics

The direct reuse strategy may be defined as: "any operation by which products or components that are not waste are used again for the same purpose for which they were conceived" (European Commission 2008). Gelbmann and Hammerl (2015) state that the performances of the directly reused product must be as good as a new one to achieve the same function while Arnette et al. (2014) assert that products have to be "good enough" to fulfil the following use. In any case, products need to be in sufficient working condition to be reused directly. Products which are reused directly are often however considered second-hand products and their components used to repair other products (Go et al. 2015) instead of becoming a product in and of themselves. This implies new products manufacturing instead of potential reuse of products. In terms of the manufacturing process, the direct reuse strategy involves already-used products' collection from the waste stream, cleaning, sorting and testing of products (Gelbmann and Hammerl 2015; Go et al. 2015). These steps make it possible to solve potential problems and ensure their well-functionality so that they can be reused directly in similar applications (Pigosso et al. 2010; Arnette et al. 2014; Gelbmann and Hammerl 2015). The remaining unsettling factor about the definition of direct reused products concerns its legal status after the first use. Some authors (Gelbmann and Hammerl 2015) insist on considering them as wastes since the European Commission (2008) no longer does this. In the latter case, the product shall ceased to be defined as such upon following different steps to be reintroduced onto the market (European Commission 2008).

The remanufacturing strategy has largely been studied over the past decades. Lund (1984) gave the first definition of remanufacturing and stated it to be: "an industrial process in which worn-out products are restored to like-new condition." This definition has been adapted by the European Commission (2015), which describes remanufacturing as "a series of manufacturing steps acting on an end-of-life part or product in order to return it to like-new or better performance, with corresponding warranty." The most important matter to appreciate here is that manufacturing processes will be needed in order to bring products back to their original state or to a better state. In other words, the remanufacturing process attempts to recover as much added-value from the original manufacture as possible (Zwolinski et al. 2006; Gray and Charter 2008). The remanufacturing process may be slightly more complex than direct reuse. The starting point for remanufacturers is to obtain from the user, the collected-used products and return them to their factories. Sundin and Bras (2005) detail seven generic process steps for the remanufacturing business: inspection, storage, cleaning, disassembly, reassembly, repair and testing. These steps—in part or in full—are found in any remanufacturing activity whatever its sector of activity.

In both direct reuse and remanufacturing strategies, the objective is the same: deliver to the market a product that is similar to the initial one and built from the initial materials. They both ensure reuse objectives, while the main difference stems from the quantity of operations needed to make the product reusable again. If the process needed to rebuild the product is mainly a cleaning process, it is considered as direct reuse. Otherwise, if the process calls for machining and more complex operations, it constitutes a remanufacturing strategy. Both strategies aim at lowering our environmental pressure. Among the different end-of-life strategies, direct reuse is said to have the best environmental and economic advantages (European Commission 2008; Arnette et al. 2014; Go et al. 2015; Gelbmann and Hammerl 2015), while remanufacturing is second (Sundin and Bras 2005; Hatcher et al. 2011; Go et al. 2015). Gray and Charter (2008) quote that the remanufacturing strategy would require 85 % less energy than manufacturing. Direct reuse should not require new high energy consuming transformations. Furthermore, they would both preserve resources, as they could be seen as "a new product avoided." Hatcher et al. (2011) furthermore add that it could be "a combination of new and reused parts." The main drawback of both strategies lies in the efficiency-in-use of the product when reused. Indeed, direct reused and remanufactured products—even if they are as-good-as-new—may be less efficient than brand-new ones due to technological evolution.

3.2 Design for Reuse

In order to evaluate the different reuse strategies, i.e. direct reuse and remanufacturing—it is important to define a common framework of analysis in line with the customary design processes.

3.2.1 Different Reuse Strategies Under a Single Framework

When designing for sustainability purposes or for the environment, it is crucial to include all the different lifecycle steps, from cradle to grave—i.e. from raw material extraction to end-of-life stages, including manufacturing and use phases (Crul and Diehl 2009). From that point, a classic description of such strategies would distinguish products characteristics from manufacturing processes, or else design from production. This may come from bygone days when design office and production planning department were two separate entities. Nowadays, with integrated design, external parameters have to be considered all along the lifecycle of the product

(Brissaud and Tichkiewitch 2000). This leads to a better organisation of the overall offer, whether it be in terms of stakeholders' relationships, value creation, value chain of the offer, or any surrounding elements. All these elements are then gathered under the business model label. A parallel has already been made in the remanufacturing literature, where Gray and Charter (2008) pinpointed these three dimensions (called spheres) and distinguished Product characteristics from manufacturing Processes and Business Model features (P.P.BM. spheres). Indeed, Sundin and Bras (2005) and Zwolinski et al. (2006) detailed product characteristics and process activities considering external factors.

The P.P.BM. spheres are considered in this paper for the purpose of structuring the design guidelines. These guidelines help designers to define product and process parameters in line with the strategy of the company. The *product* area covers the product itself and its components. Their main characteristics are defined in order to distinguish products from different EoL strategies. The *process* concerns the different steps put in place in order to deliver the products and assign their respective characteristics. The Business Model defines the global strategy for delivering the product and its organisation. Each of these three spheres entails specific characteristics defined from literature in Bauer et al. (2016) and recalled in Tables 1, 2 and 3.

3.2.2 How to Design These Kinds of Products?

Design processes have largely been studied in the literature (Tomiyama et al. 2009). Design tools and methods have been well-known for years and many improvements have already been made, especially with integrated design (Brissaud and Tichkiewitch 2000). Indeed, designing a product implies the interaction from multiple areas of expertise in a single company. In that process, gathering the different actors from the early stages would facilitate the integration of the different constraints, whether they were linked to the product, the process, or the business model. From that point, the design process follows different steps to progress from the product idea to the product retirement (see Fig. 2).

Although they follow a reuse strategy at their end of use, to-be-reused products need to be considered like any other manufactured ones in the first place, so that the design phases between the two would not change much (Gray and Charter 2008). Despite that, the key issue for to-be-reused products lies in integrating the required parameters that are designed to ensure the end-of-life strategy. To be set up efficiently, they have to be integrated from the early design stages (Gray and Charter 2008). Hence, reuse can be seen as a classic integrated design, with specific attention to end-of-life parameters.

Table 1 Guidelines: process sphere (classified by characteristics)

Characteristics	Guidelines for reuse strategies	Guidelines for Remanufacturing strategy only
Stable process	Standardise and use common tools Reduce the diversity of components Reduce the variation in cores	
Inspection & Sorting	Minimise inspection time Mark inspection points clearly Minimise the number of different materials Use standard components	
Cleaning	Avoid components that can be damaged during cleaning process Minimise geometric features harming cleaning process Identify components requiring similar cleaning processes	Facilitate access to the cleaning process Ensure marking on product can survive cleaning process
Dis/Re-assembly	Avoid permanent fasteners that require destructive removal Increase corrosion resistance of fasteners Reduce the total number of fasteners Reduce the number of press-fits Standardise and use common fasteners (type and size)	Minimise disassembly and reassemblyt ime Arrange parts and components to facilitate assembly, especially the ones that are easily prone to damage Use assembly techniques that allow easy access to inspection points Use assembly techniques that allow upgrade Use assembly techniques that will withstand overall remanufacturing processes but that will not allow for damage to components that have the potential to be reused/ remanufactured Use robust materials to ensure assembly operations
Storage	Ensure no damage during storage	
Remanufacturing		Standardise and use common processes
Testing	Minimise thenumber of tests Reduce test complexity Standardise tests Reduce the number of tests at the level required Facilitate tests of components Provide testing documentation	

3.3 Main Guidelines for Design for Direct Reuse and Remanufacturing

The reuse literature is already overflowing with design guidelines for facilitating the adoption of direct reuse and remanufacturing strategies (Ijomah 2009; Arnette et al. 2014; Go et al. 2015). In the same manner, three spheres have been proposed (Bauer et al. 2016) to characterise end-of-life strategies. The categorisation of

Table 2 Guidelines: product sphere (classified by characteristics)

Characteristics	Guidelines for reuse strategies	Guidelines for Remanufacturing strategy only
Reliable product	Select reliable materials Select reliable components	
Durable product	Select durable materials Select durable and robust components Prevent core damage Prevent part and surfaces against external environment Avoid components that can be damaged during cleaning process	Avoid components that can be damaged during inspection process Avoid components that can be damaged during disassembly process Avoid components that can be damaged during refurbishment process
Functional prob.		
High initial cost		
Efficient product		
Modularity / Upgradability		Standardise and use common materials, components and fasteners Use modular parts and components thus reducing complexity of disassembly because types of assembly techniques are reduced Structure the product and parts to facilitate ease of upgrade
Physical elements		Avoid permanent fasteners that require destructive removal Increase corrosion resistance of fasteners Standardise and use common fasteners (type and size) Reduce the total number of parts, components, fasteners, press-fits and joints Specify materials and forms appropriate for repetitive manufacturing
Stable technology	Standardise and use common materials, components and fasteners Standardise and use common interfaces Design reusable parts and components Facilitate access to components Facilitate switch of damaged components	
Documentation	Provide readable labels, text, and barcodes that do not wear off during the product's service life Provide good documentation of specifications, clear installation manuals and testing documentation Provide clear information about product, parts, components and materials Set up sacrificial parts to give an indication of the components' state of life	

Table 3 Guidelines: business model sphere (classified by characteristics)

Characteristics	Guidelines for reuse strategies	Guidelines for Remanufacturing strategy only
Ease of reuse	Verify the market acceptance of the offer Determine the internal skills needed	Reduce the rejection of remanufactured products
Ease of supply	Embed mechanisms into the product to ensure the return of cores Facilitate collection of core parts Facilitate Reverse logistics	
Economic motivation		
User profile		
Remanufacturing reason		
Partnership		
Legislation		
Environmental gains	Avoid toxic materials Determine the cleaner production anduse	

Fig. 2 Common design stages in product development

design guidelines according to one of the P.P.BM. sphere and then to the closest characteristic it would be linked to, is what is proposed here. Designers are therein provided with the guidance necessary for identifying which rule would lead to which characteristic. Some characteristics are created or renamed when the initial ones are not relevant enough for a design activity.

The two specific reuse strategies—direct reuse and remanufacturing, their characteristics and design guidelines are classified in Tables 1, 2 and 3. Guidelines dedicated to both direct reuse and remanufacturing were grouped together in one column labelled 'Guidelines for reuse strategies', while the ones specific to remanufacturing were separated in the right-hand side column. Table 1 thus clusters characteristics and guidelines addressing the process Table 2 then gathers the different characteristics and guidelines connected to the product. The principal elements are related with direct product characteristics, such as durability and reliability, and physical elements facilitating the strategy—e.g. fasteners, parts.... The main process steps are recalled and specific guidance is provided. Finally, Table 3 covers the business model characteristics and guidelines. It is mainly a matter of organisation and reverse logistics.

Two points immediately stand out for careful discussion. First, it appears that some characteristics do not have any concomitant guideline. The reasons are that none of them has been identified in literature or due to the fact that the guideline was closer to another characteristic. The knowledge corpus will be increased with literature progress. The second point concerns the repartition of the guidelines. It appeared that all the guidelines related to direct reuse strategy were included in the remanufacturing strategy (grouped together in the column labelled '*Guidelines for reuse strategies*'). Nevertheless, some have only been identified in DfRem literature. This seems logical, however, as, the major difference between both is that remanufacturing implies more remanufacturing before the product could go back to the market. This is noticeable in the Tables 1, 2 and 3: all specific remanufacturing guidelines are directly or indirectly related to the remanufacturing process steps.

3.4   *Discussion*

The characteristics have been highlighted and organised according to the P.P.BM. spheres. Design guidelines from literature were then linked to the most relevant reuse characteristic. If everybody agrees on the end goal of maintaining a high level circular economy, the applications are not as numerous as expected (Gelbmann and Hammerl 2015). Reused products may not yet be well-accepted on the market (Arnette et al. 2014), nor are design guidelines practical enough for each particular product.

The primary difficulties in implementing the reuse strategies remain. One key parameter concerns the reverse logistic chain, hitherto not well addressed as it mainly depends on company decision-making (Hatcher et al. 2011; Go et al. 2015). Indeed, the crucial step is to retrieve already-used products in pursuit of ensuring direct reuse or remanufacturing. This issue has to be defined from the design stages (Go et al. 2015). That is, the company needs to know where the retired products will be, how to get them back, and how to set up the logistics for bringing them back to the company or to another defined point (Gelbmann and Hammerl 2015; Go et al. 2015). These steps may rely on partnerships (Gelbmann and Hammerl 2015). The second point is related to the difficulty in putting the strategies in place *a posteriori*, after the products have been designed and *lived* (Hatcher et al. 2011). The use of the precedent design guidelines may allow for partial avoidance of such problems, or at minimum, for identification of the weak points ahead.

The limits of the design guidelines for reuse strategies need also to be highlighted. First of all, characteristic to all guidelines is that they tend to be rather generic, which means they should be applicable to most of the products. Designers need to adapt them to the case at hand, yet the resulting specifications may conflict with the guidelines traditionally used in the domain. Secondly, some of the characteristics that have been highlighted in each sphere do not contain any guidelines either for direct reuse or remanufacturing. Two main reasons can be outlined here. Number 1: the characteristic is mainly related to the company strategy and its motivation for this kind of business—e.g. economic motivation, favouring legislation. All the same, no generic guideline is applicable as it is related to the company itself. Number 2: the characteristic is inherent to the product itself and is more related to product specifications than guidelines—e.g. high initial cost, efficient product. Guidelines, company specifications and product specifications are complementary and thus, it does not matter in what manner they find their way to the designer. In practice, when using DfdR and DfRem guidelines, a risk arises that designers follow the guidelines without integrating the initial product and process specifications and therein miss out on some crucial points. Guidelines are set up to facilitate the designer's job according to previous studies. Yet, every product is distinct from the others, so that requiring specific parameters may make one guideline irrelevant and may thus not apply.

4 The Repurposing Strategy

A rising EoL strategy in literature concerns "repurposing". Repurposing is a third reuse end-of-life strategy that complements the two previous ones. Much like other reuse strategies, repurposing allows for retention of added-value in used products.

4.1 Limits of Direct Reuse and Remanufacturing Strategies

Current reuse strategies—i.e. direct reuse and remanufacturing—aim at and succeed in preserving a part of the added-value of used products in the manufacturing of new products. The reuse process can be seen in three main issues (Fig. 3). The limits of each of them are analysed for the purpose of extracting the orientations for a complementary strategy that would increase the quantity of reused products.

The reuse strategy is a manufacturing strategy driven by market conditions. The assumption in direct reuse and remanufacturing is that the new product must at least offer the same levels of performances and of customers' satisfaction than the old product. The market can be limited by the number of like-new products that can be absorbed by the customers. The market must furthermore be open for new products. Opening the market involves upgrading or repurposing. Upgraded products are products of the initial family where performances and functions are different. Repurposed products are products that are sold for a different purpose and belong to a different product family. For example, electric vehicles' batteries can be recomposed to be reused in stationary applications.

The existence of the transformation process depends on the technical feasibility *(can the process push the product to the initial performance?)*, the environmental performance *(is the reused process greener than the initial one?)* and the economic concern *(can the value chain be profitable?)*. Because products are very often designed without any objective of reusing them, they cannot be disassembled

Fig. 3 The reuse end-of-life main process

without damage and, consequently, cannot be reused. It is thus clear that design is a very important phase to improve upon. Yet there are also remanufacturing processes that cannot give back the initial performances to the product. It is clearly the case today for batteries of electric vehicles that cannot be remanufactured for the simple reason that the technology is unable to recover the initial performance at a reasonable cost (Beverungen et al. 2016). The question of what to do with the stock of such batteries is an open issue.

The collection of already-used products depends on their quality *(does the core retain the quality for the expected performance?)* and quantity *(are there enough collected used products to make the business profitable?)*. Quality issues could sometimes be overlooked if the question was raised of finding new applications where technical performances are not the key issue. Quantity depends on the efficiency of the collection process and the capacity of the market to absorb more products. Alongside the economic issue, the environmental issue of waste management can likewise figure in as a significant driver of the business.

Let us explain the concept with the example of electric vehicle batteries, currently under discussion in the literature. It starts with two claims: in a few years' time, the issue of waste management will be crucial because the performance of a battery cannot be recovered by technology, while the market of stationary applications calling for batteries is however exploding. The idea is to couple both claims and see whether electric vehicle batteries, no longer efficient enough for mobile applications, can be reused after transformation in stationary applications like lighting and housing. Idjis (2015) studied a recovery network for end-of-life electric vehicle batteries from "a technical-economic, organizational and prospective perspective." He identified the business model elements (the economic viability; legal requirements) that enable the repurposing of a company to manage reverse logistics for core supply, to rely on partnerships, and assessed the effective quantity of batteries for repurposing into stationary applications as well as the properties at the end-of-use. Beverungen et al. (2016) identified and validated with experts the functional and non-functional requirements for repurposed batteries from EV to stationary applications. Based on a battery expert interview and literature (Ahmadi et al. 2014; Bauer et al. 2016; Beverungen et al. 2016), the repurposing process seems to include the same steps as reuse strategies: inspection and sorting, cleaning, dis-/re-assembly, storage, repurposing operations and testing. The repurposing step would mainly rely on reconfiguring the different components and sub-assemblies of the products and include a few product developments in order to then fulfil new requirements or connect the components in the new fashion.

4.2   Repurposing: Definition and Advantages

Repurposing is a reuse end-of-life strategy that aims at preserving added-value of used products by reusing them in different applications and fields and in so doing, get around the remanufacturing and direct reuse strategies by targeting new markets.

Repurposing aims at maintaining high added-value products on the market as long as possible, to ultimately delay recycling or disposal. This strategy does not replace direct reuse or remanufacturing, but nevertheless fills a gap when these two last options are not applicable. No market cannibalisation may take place, as, the applications are distinct. This strategy should complete the list of reuse strategies and contributes to extended producer responsibility in the whole environmental consciousness equation (European Commission 2008). Company responsibility at the end of the first end-of-usage is transferred to the second life of the products. It could be done in as many cycles as possible until being transferred to the material recycling process. When the repurposing is properly implemented, the strategy is determined to be more environmentally friendly and less cost effective than manufacturing products from raw materials. The research only still has to prove in which conditions this performance may be present.

The repurposing process is close to a remanufacturing one (Fig. 4). The same types of operations are necessary, even when the combinations of parts are larger. The main difference is that the diagnostic phase on the quality of the used products collected (the product health) must be much more detailed and very intelligent in pursuit of orienting the core to the most adapted transformation process. Another difference of course lies in the technology for transforming the used product into a totally different product that must be developed, which then turns out to be easier in terms of repurposing. This strategy holds great potential for personalising new products. The principle that the performance criteria may evolve from one use to another points to real opportunity in that realm.

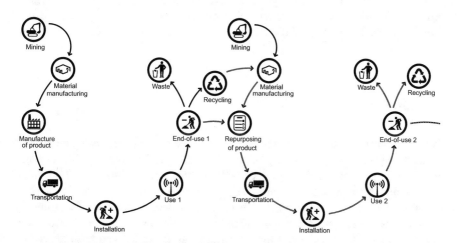

Fig. 4 Product lifecycle for repurposing, the end-of-life strategy

4.3   Short Discussion on Design for Repurposing

Design for repurposing represents a completely new issue. If it seems adapted to benefit from the guidelines for reuse strategies presented above, then the perspective of the design becomes totally different, meaning that the design drivers should be re-conceptualised.

The main discussion is on determining whether the best design strategy is to design the new products from a classical design process where the constraints of input elements are new (the collected parts and materials) but known, or to design products from scratch that would have several lives in different applications. The former calls for research in defining the specifications of a repurposed product along with the design rules for transforming a product with a repurposing approach. The latter seems to be much more optimal, but the uncertainty attached to the future of the product is so high that anticipating the actual usages and the time of the first use, yields only clues about short life products. Furthermore, additional difficulty stems from the number of different applications necessary for consideration before the original design phase. The new design approach, in the both cases, should include an objective of monitoring successive lives of the product in order to help decide on the parameters of the next life once the time comes.

The literature has commenced, with Beverungen et al. (2016) and Bauer et al. (2016) already proposing some characteristics of repurposed products and repurposing production systems. The repurposed system has to be durable and reliable, which means that few instances of breaking should happen during its lifetime, while its performance should be possible to predict. Safety issues must also be addressed differently, i.e. extra life products need to consider safety as a key element for the consumer. They highlight that modularity and standardization would help to that effect. In the end, however, the principles are the same: physical characteristics of products should facilitate the repurposing process. All these points have not yet been addressed in full in the design literature and further investigations are therefore needed.

5   Conclusion

Design for direct reuse and remanufacturing, the end of use strategies with the most added-value retained from used products, have already become a reality in companies and are in demand by society with sustainability ambitions. While direct reuse is mainly a logistics and control issue, remanufacturing aims at getting back to the initial performances of products. These two strategies have been fully examined in studies of the last years and their main characteristics were presented according to three spheres: product dimension, manufacturing processes and business model features (P.P.BM.). The design guidelines were collected and classified for an easy use by designers.

To open minds, a valuable strategy for reusing products in different applications than the initial ones were designed for is proposed: repurposing. The concept is clarified and the main issues for the design process have been highlighted. These pursuits are promising but need investigation to find the conditions for successful deployment.

References

Ahmadi, Leila, Michael Fowler, Steven B. Young, Roydon A. Fraser, Ben Gaffney, and Sean B. Walker. 2014. Energy efficiency of Li-Ion battery packs re-used in stationary power applications. *Sustainable Energy Technologies and Assessments* 8(December): 9–17. doi:10.1016/j.seta.2014.06.006.

Arnette, Andrew N., Barry L. Brewer, and Tyler Choal. 2014. Design for sustainability (DFS): The intersection of supply chain and environment. *Journal of Cleaner Production* 83(November): 374–390. doi:10.1016/j.jclepro.2014.07.021.

Bauer, Tom, Guillaume Mandil, Élise Naveaux, and Peggy Zwolinski. 2016. Lifespan extension for environmental benefits: A new concept of products with several distinct usage phases. In *Procedia CIRP*, 47:430–435. Product-Service Systems across Life Cycle. doi:10.1016/j.procir.2016.03.079.

Beverungen, Daniel, Sebastian Bräuer, Florian Plenter, Benjamin Klör, and Markus Monhof. 2016. Ensembles of context and form for repurposing electric vehicle batteries: An exploratory study. *Computer Science—Research and Development*, July, 1–15. doi:10.1007/s00450-016-0306-7.

Brissaud, Daniel, and Serge Tichkiewitch. 2000. Innovation and manufacturability analysis in an integrated design context. *Computers in Industry* 43(2): 111–121. doi:10.1016/S0166-3615(00)00061-0.

Crul, Marcel, and Jan Carel Diehl. 2009. *Design for sustainability: A step by step approach.* UNEP, United Nation Publications. http://www.d4s-sbs.org/.

European Commission. 2008. *Directive 2008/98/EC of the European Parliament and of the Council of 19 November 2008 on Waste and Repealing Certain Directives.* http://eur-lex.europa.eu/legal-content/EN/TXT/?uri=celex:32008L0098.

European Commission. 2015. *Communication from the commission to the European Parliament, the Council, the European Economic and Social Committee and the Committee of the Regions. Closing the Loop—An EU Action Plan for the Circular Economy.* Bruxelles, Belgique. http://ec.europa.eu/environment/circular-economy/.

Gelbmann, Ulrike, and Barbara Hammerl. 2015. Integrative re-use systems as innovative business models for devising sustainable product–service-systems. *Journal of Cleaner Production* Special Volume: Why have 'Sustainable Product-Service Systems' not been widely implemented?, 97 (June): 50–60. doi:10.1016/j.jclepro.2014.01.104.

Global Reporting Initiative. 2013. G4 sustainability reporting guidelines—reporting principles and standard disclosures. Global Reporting Initiative. https://www.globalreporting.org/resourcelibrary/GRIG4-Part1-Reporting-Principles-and-Standard-Disclosures.pdf.

Go, T.F., D.A. Wahab, and H. Hishamuddin. 2015. Multiple generation life-cycles for product sustainability: The way forward. *Journal of Cleaner Production* 95(May): 16–29. doi:10.1016/j.jclepro.2015.02.065.

Goodall, Paul, Emma Rosamond, and Jenifer Harding. 2014. A review of the state of the art in tools and techniques used to evaluate remanufacturing feasibility. *Journal of Cleaner Production* 81(October): 1–15. doi:10.1016/j.jclepro.2014.06.014.

Gray, Casper, and Martin Charter. 2008. Remanufacturing and product design. *International Journal of Product Development* 6(3–4): 375–392.

Hatcher, G.D., W.L. Ijomah, and J.F.C. Windmill. 2011. Design for remanufacture: A literature review and future research needs. *Journal of Cleaner Production* 19(17–18): 2004–2014. doi:10.1016/j.jclepro.2011.06.019.

Idjis, Hakim. 2015. La filière de valorisation des batteries de véhicules électriques en fin de vie : contribution à la modélisation d'un système organisationnel complexe en émergence. Ph.D. thesis, Université Paris-Saclay. https://tel.archives-ouvertes.fr/tel-01243863/document.

Ijomah, Winifred L. 2009. Addressing decision making for remanufacturing operations and design-for-remanufacture. *International Journal of Sustainable Engineering* 2(2): 91–102. doi:10.1080/19397030902953080.

ISO. 2002. *Integrating environmental aspects into product design and development. ISO/TR 14062:2002.* http://www.iso.org/iso/home/store/catalogue_tc/catalogue_detail.htm?csnumber= 33020.

Lund, Robert T. 1984. Remanufacturing. *Technology Review* 87(2): 19–29.

Pigosso, Daniela C.A., Evelyn T. Zanette, Américo Guelere Filho, Aldo R. Ometto, and Henrique Rozenfeld. 2010. Ecodesign methods focused on remanufacturing. The roles of cleaner production in the sustainable development of modern societies. *Journal of Cleaner Production* 18(1): 21–31. doi:10.1016/j.jclepro.2009.09.005.

Sundin, Erik, and Bert Bras. 2005. Making functional sales environmentally and economically beneficial through product remanufacturing. *Journal of Cleaner Production* 13(9): 913–925. doi:10.1016/j.jclepro.2004.04.006.

Tomiyama, T., P. Gu, Y. Jin, D. Lutters, Ch. Kind, and F. Kimura. 2009. Design methodologies: Industrial and educational applications. *CIRP Annals—Manufacturing Technology* 58(2): 543–565. doi:10.1016/j.cirp.2009.09.003.

Zhang, Feng. 2014. *Intégration Des Considérations Environnementales En Entreprise : Une Approche Systémique Pour La Mise En Place de Feuilles de Routes.* Grenoble, France: Université de Grenoble. http://www.theses.fr/s95811.

Zwolinski, Peggy, Miguel-Angel Lopez-Ontiveros, and Daniel Brissaud. 2006. Integrated design of remanufacturable products based on product profiles. *Journal of Cleaner Production* 14(15–16): 1333–1345. doi:10.1016/j.jclepro.2005.11.028.

Target-Driven Sustainable Product Development

Tom Buchert, Anne Pförtner and Rainer Stark

Abstract Figuring in sustainability in product development requires a profound understanding of the cause and effect of engineering decisions along the full spectrum of the product lifecycle and the triple bottomline of sustainability. Sustainability design targets can contribute to mitigating the complexity involved, by means of a formalised problem description. This article discusses how sustainability design targets can be defined and presents methods for systematically implementing these targets into the design process. To that end, different means of decision support mechanisms are presented. They comprise (a) use cases of target breakdowns in subsystems, (b) systematic reduction of solution space and (c) assistance in design activities to ensure achievement of sustainability design targets. This paper explains how interfaces to engineering tools such as Computer Aided Design/Engineering (CAD/CAE) or Product Data/Lifecycle Management (PDM/PLM) can be put in place to make the process of retrieving information and providing decision support more seamless.

Keywords Decision support · Sustainable design · Product development · Sustainability targets

1 Challenges in Sustainable Product Development

The topic of Sustainable Product Development (SPD) has been discussed in academic research since the early nineties with a strong focus on the environmental perspective (e.g. by Allenby 1991). In this context, numerous approaches have been developed, while some success-stories, e.g. the diffusion of LCA into industrial

T. Buchert (✉) · A. Pförtner
Technische Universität Berlin, Berlin, Germany
e-mail: Tom.buchert@tu-berlin.de

R. Stark
Chair of Industrial Information Technology, Institute for Machine-tools and Factory Management, Technische Universität Berlin, Berlin, Germany

© The Author(s) 2017
R. Stark et al. (eds.), *Sustainable Manufacturing*, Sustainable Production, Life Cycle Engineering and Management, DOI 10.1007/978-3-319-48514-0_9

129

practice (Kara et al. 2014), have been achieved. However, nearly thirty years after the sustainability debate emerged, industrial production remains far from being sustainable [e.g. in the sense of exceeding planetary boundaries (Steffen et al. 2015)]. This insight leads to the question of what specific challenges need to be overcome in product design to improve the overall situation.

From a practical perspective, a range of factors influence the successful implementation process of SPD, such as:

- personal motivation of actors (e.g. incentives for fostering sustainability integration into product design),
- available resources (e.g. time budget for SPD method application) or
- lock-in effects (e.g. existing contracts with suppliers of unsustainable materials).

While these practical barriers can be solved by appropriate managerial oversight in the respective companies, great potential remains untapped in the research on SPD. A major challenge in this context is to find solutions for decreasing the complexity attached to SPD decision-making. Yet a sustainable design can only be achieved if design engineers develop subsystems in accordance with their influence on the triple bottomline (economic, environmental and social sustainability) at each and every step along the way of the entire product lifecycle (see Fig. 1). One approach for coping with this complexity is to break the problem down to smaller sub-problems which are easier to handle (problem modules). Figure 1 gives an example of which modules can be considered in the context of SPD (e.g. environmental impacts of electronic recycling).

Nevertheless, this reductionist approach may not prove to be sufficient due to the diverse interrelations between problem modules (e.g. better recyclable electronics may lead to economic problems in production). A key task of SPD research

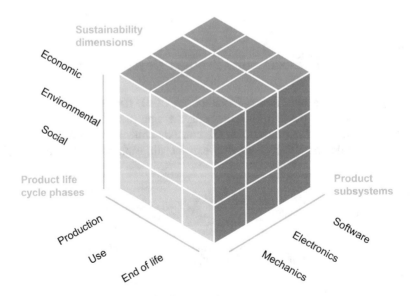

Fig. 1 Modules of sustainable product development problems

therefore lies in enabling engineers to anticipate these dependencies by means of methodological guidance as well as by enhanced knowledge and information supply. Thus, rather than searching for solutions to specific problem modules, this article will focus on providing novels mechanisms for increasing the transparency of decision-making.

2 Methods for Supporting Decision-Making in SPD

A wide variety of approaches for supporting decision-making in SPD have been developed in the last years. Baumann et al. (2002) classify existing approaches for environmental sustainability into six categories which still apply today in the field of SPD:

Analytical tools try to quantify the impact on the three dimensions of sustainability with varying precision. Life Cycle Sustainability Assessment as a combination of Life Cycle Assessment, Life Cycle Costing and Social Life Cycle assessment (Neugebauer et al. 2015) is utilised for more accurate estimations in later design phases, at which point plenty of information about the product is available. More simplified approaches (e.g. by Collado-Ruiz and Ostad-Ahmad-Ghorabi 2013) can be utilised in earlier phases as a form of heuristic prediction of impacts.

Checklists and guidelines provide best practices for guiding engineers along in the design process. They can be utilised in the early phases but are less helpful for decision-making for specific design problems. In the environmental realm, exhaustive collections of design guidelines have long since existed (Telenko et al. 2016). Guidelines for sustainable design are scarce. The most mature approach is based on a modular set of guiding questions which are also referred to as "templates" (Ny et al. 2008).

Rating and ranking tools provide possibilities for the simple but structured comparison of different solution options, based mostly on qualitative or semi-quantitative evaluation (see for example Shuaib et al. 2014).

Organizing tools furthermore help structure the design process by involving multiple stakeholders in the form of workshops or structured interviews.

Software and expert systems assist in applying methods by automating certain steps of the method application or by simplifying the process of researching for information through databases (e.g. LCA software such as GaBi.[1] Furthermore, IT-support of this kind has the potential to enable one's own methodological approaches like the Eco-Pas software tool by Duflou and Dewulf (2005). The latest approach for IT-based decision support is the integration of SPD methods in engineering tools like CAD (e.g. Solidworks Sustainability Pro[2] and in PDM systems (Ciroth et al. 2013). Nevertheless, these approaches are limited to the

---

[1]https://www.thinkstep.com/software/gabi-lca/.

[2]http://www.solidworks.de/sustainability/.

assessment of the current design progress and the relative comparison to a reference design without actual guidance. Furthermore, the underlying impact model and the dependencies between engineering decisions as well as the sustainability impact have all yet to be made transparent to the engineer. In particular, trade-offs between the sustainability dimensions are not being intensively researched since most of available methods focus on the environmental sustainability perspective. The following three characteristics summarise insights on methods for sustainable product development (see Buchert et al. 2014 and Sect. 4.3):

- Existing methods focus on assessment. There is a lack of engineering approaches that assist engineers in the form of offering support for design synthesis. Guidelines can be utilised for synthesis but are often not sufficient for addressing a specific design problem.
- Availability of information in conceptual design is usually scarce. Analytic approaches require a lot of information and are therefore only utilised once major decisions have already been made.
- Insufficient transparency on system relations between product design decisions, sustainability impacts and product life cycle stages prohibits a systematic examination of the specific trade-offs and side-effects attached to engineering decisions.

3 Integration of Sustainability Targets into the Design Process

The complexity of cause and effect chains presents a major challenge for judgment calls in sustainable product development. One favourable way of reducing the complexity factor in the whole process lies in defining targets which specify the most relevant influencing factors for the problem and which provide a basis from which to develop further decision-making models (Bretzke 1980). Hence, it needs to be clarified how "sustainability design targets" can be formulated in a complimentary fashion to basic technical or functional requirements. A starting point for addressing this problem is to pinpoint the causal relations between engineering decisions and sustainability impact. This is achieved by classifying different types of information to different categories in a fixed order illustrated in Fig. 2. The categories and their respective relationships will be introduced in the following paragraphs. A more detailed description can be found in (Buchert et al. 2016).

The scheme developed is based on the separation of product characteristics (I) and properties (II) as defined by Weber et al. (2003) in the scope of their "Property Driven Product Development (PDD)" approach. Category (III) refers to the sustainability impact of a product on various stakeholders such as the environment, health aspects of employees and customers as well as the financial stability of the company (III). In order to connect the design engineering perspective (I and II) with the sustainability impact view (III), the category *product properties*

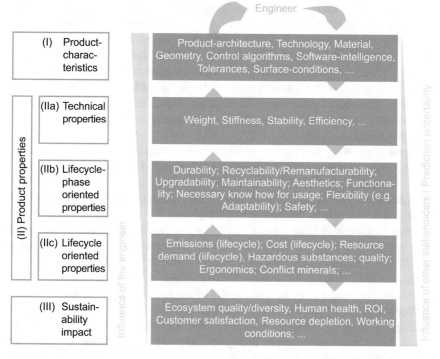

Fig. 2 Scheme for linking design decisions to sustainability impact (Buchert et al. 2016)

was subdivided into three subclasses, each of which takes the perspective of the product life cycle (IIa–IIc) into consideration. Category IIa consists of technical properties that are defined directly as a result of engineering decisions for product characteristics. The definition of the characteristic's material and geometry defines, for example, the technical property weight. When technical properties are combined and enriched with influences from outside the system, boundary lifecycle-phase oriented properties are determined (IIb). The prediction of durability in the usage phase of a pedelec frame is, for example, based on technical properties such as tensile strength or stiffness, but also relies on user behaviour. This type of property defines how a product interacts with its surrounding systems within specific life-cycle phases (e.g. durability, remanufacturability). If all effects of these interactions are aggregated along the product lifecycle, (e.g. in terms of cost or emissions) lifecycle-oriented properties are then derived (IIc). Lifecycle oriented properties can be understood analogous to the term Lifecycle Inventory which is used in the context of Life Cycle Assessment to evaluate environmental sustainability.

By analysing the complete scheme, it becomes evident that the influence of engineering decisions decreases with every level, since other actors in product creation (e.g. sourcing) likewise have a significant influence on overall product sustainability impact. Furthermore, external influences (e.g. user behaviour) may deviate from assumptions stemming from the design process and therefore increase uncertainty of predictions for the whole lifecycle of a product. One conclusion from this analysis is that targets on impact level are less appropriate for design engineers since they are not trained to relate their actions properly to impact indicators. Hence, sustainability experts need to be involved in the design process, which serves to make the most critical lifecycle-oriented properties and the most significant lifecycle phases for engineering target definition identifiable. In addition to sustainability experts, other company roles may also define relevant targets in a sustainability context, in particular from an economic perspective (e.g. product or quality managers).

Figure 3 provides a reference framework for integrating sustainability targets into the design process by differentiating between different stakeholders involved and by identifying challenges for successful target integration. Potential for decision support in the other direction is identified with this framework. Respective challenges are introduced in the following paragraphs.

Once sustainability targets are defined by the respective experts, they then have to be broken down into technical subsystems or assemblies by system engineers (see Fig. 3). This step poses a special challenge, since it is not clear how narrowly a target should be formulated in order to be effective in the sense of sustainability improvement. It can be argued which level of the scheme shown in Fig. 2 is most appropriate for which respective purpose. The more specifically the target is defined (e.g. on the level of technical properties such as weight), the less opportunities remain for domain engineers to find a creative solution to foster sustainability performance. Furthermore, unintended side effects can occur since the domain

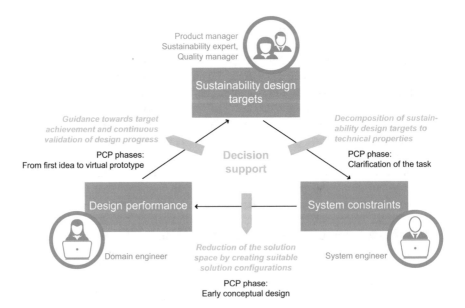

Fig. 3 Framework for decision support based on sustainability design targets

engineers may not be informed about the intended effect of the target in terms of sustainability improvement (e.g. changing to a lighter material to save fuel consumption of a car may shift the environmental burden to material production). In this context, how exactly targets should be allocated to subsystems for establishing the basis for a sustainable solution configuration also needs to be evaluated.

Another challenge which needs to be overcome to properly address sustainability targets in the design process, lies in the identification of sustainable and technically feasible configurations of system elements and inherent product characteristics. This task is troublesome since multiple configurations are possible, and it also needs to be determined which components can be reused and where new developments are necessary. This reduction of the solution space decreases the effort for further design activities and therefore increases efficiency and effectiveness of the design process.

Domain engineers then finally develop suitable solutions according to the given requirements. In that pursuit, it is necessary to evaluate whether the current design process and estimated product performance in different PCP stages are compliant with given sustainability design targets. Furthermore, providing specific advisory tips towards achieving these targets can be beneficial. Hence, a range of activities might be necessary, such as, ideation for new and more sustainable products, comparison of solution concepts, and/or final solution assessment. A broad set of methods has been developed for assisting in these tasks. The challenge therein lies in selecting the right method for each and every task along the way in the product development process.

The challenges described are also summarised in Fig. 3 and are viewed by the authors a handy framework of reference for implementing sustainability targets in the design process. Decision support tools can play a viable role in overcoming these challenges by fostering transparency on sustainability cause and effects and by increasing the availability of information for the engineer. New approaches for decision support to that effect are therefore presented in the following chapters addressing these aspects.

4 Decision Support for Integrating Sustainability Design Targets

This section introduces three concepts for addressing the challenges for integrating sustainability targets into the design process. The respective approaches are explained based on the example of a pedelec (i.e. an electric bicycle) redesign project. Exemplary questions raised within the scope of this project are illustrated in Fig. 4 and will provide use cases for decision support mechanisms which have been developed.

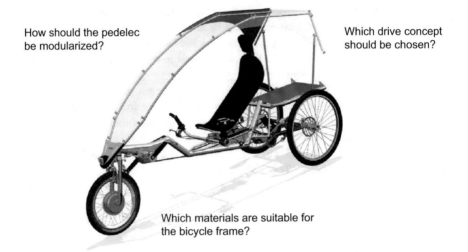

How should the pedelec
be modularized?

Which drive concept
should be chosen?

Which materials are suitable for
the bicycle frame?

Fig. 4 Exemplary engineering decisions with an influence on pedelec sustainability

4.1 Breakdown of Sustainability Targets for Product Architecture Decisions

Proper breakdown of sustainability targets from desired impacts to technical influencing-factors introduces an array of hurdles for design teams. Yet defining targets at the system level and for respective subsystems can be seen as one of the most crucial tasks in the design process, since the basis for implementing engineering strategies is defined in this step. In many companies, heuristics are followed to define their strategies. Automotive companies choose, for example, "lightweight design" to reduce fuel consumption of their vehicles. The problem with heuristics is that they are often formulated for one specific target without considering side-effects and concomitant relations with other company targets. Thus, it can be helpful to give an overview of which options are available to achieve targets or, on the flip-side, to see which indicators can be affected by design changes. A good example of a missing link in cause and effect chains can be found when considering decision-making on product architecture. The majority of companies modularise their products to limit internal complexity, to decrease their time to market, and to increase external variance for customers at the same time (Gleisberg et al. 2012). Nevertheless, other relevant targets have to be considered, such as flexibility of the product to allow multiple product use-cases and disassembly to provide reuse opportunities or simplification of maintainability.

In order to increase the transparency of relations between modularisation decisions and sustainability targets, a qualitative concept map was developed. An extract of the map is displayed in Fig. 5. The full map contains 77 modularisation drivers (i.e. targets and sub-targets) and 44 modularisation metrics. The map is structured from top to bottom regarding the information categories identified in Fig. 2 and the type of decisions addressed (from strategic to tactic and from tactic to operational level). The grey boxes visualised in Fig. 5 mark one possible way through the map starting with sustainability targets on impact level at the top. This particular way is explained for the example of setting targets for a pedelec architecture definition. At impact level (III), different sustainability indicators may be relevant for a pedelec. In this example, climate change and customer value were chosen as important impact categories. In practice, the selection of indicators relies on legal requirements as well as on company strategy, which may also include

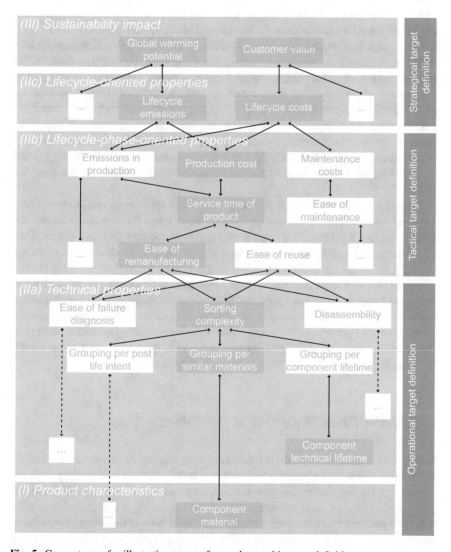

Fig. 5 Concept map for illustrating targets for product architecture definition

voluntary agreements. Customer value relies on the total cost of ownership (life-cycle cost) of the pedelec. Yet, there are also other factors to consider, such as functionality, which can be enhanced by upgradeability of the pedelec (e.g. with a stronger motor or an additional roof). The environmental impact category Climate Change is determined by total greenhouse gas emissions (GHG) along the pedelec lifecycle.

To reduce total emissions, the production phase of the pedelec should be considered since it contributes almost half of the total GHG emissions of a pedelec (Neugebauer et al. 2013). The most important contribution of modularisation at decreasing GHG emissions in production is to increase the time the product can be used (service time), since a longer utilisation period ultimately decreases the amount of products which have to be manufactured. If less products need manufacturing, absolute production cost likewise decreases. Furthermore, remanufacturing or reusing are possible measures for increasing the service time of the product. Both End of Life (EOL) options can be fostered by increasing the ease of disassembly or by grouping components in such a way that the sorting of components can be improved upon (e.g. by clustering components with same materials).

In contrast to other product characteristics, like material or geometry decisions, sustainability targets cannot be broken down to the individual component level (e.g. a targeted efficiency of a motor). Targets for modularisation can only be formulated on a system level since modularisation considers how different components are organised.

By going through the map, it should be noted that the strategic paths chosen may also lead to side effects. Increasing service time may, for example, impact the business model by decreasing sales revenue due to the fact that less products are sold. Furthermore, production could turn out to be less efficient, leading to the necessity of downsizing the production system. Smaller production may lead to personnel shifts, layoffs etc. Due to this multitude of effects, it can be difficult to find a suitable system boundary for strategic modularisation decision-making. Furthermore, missing quantification of relations between targets and modularisation metrics poses a barrier toward the quantitative assessment of modularisation effects. For enhanced decision-making in support roles, new quantified models for modularisation impact will thus have to figure into play (see Sect. 5).

4.2   Model Based Reduction of the Solution Space

Targets which are broken down and formulated as constraints can be used to reduce the solution space and eliminate the design solutions that do not comply with the defined constraints. Calculating the fulfilment of constraints for possible solutions manually is however time-consuming. Each option for all variable characteristics (e.g. each material and geometric parameters) would have to be assessed in order to determine all the viable solutions. If relations between a choice of characteristic and constraints associated with a target are formalised on a quantitative basis, viable solutions can be calculated automatically. Consequently, a formalised model expands the option pool for considering a high amount of configurations and multiple targets. Configuration options from predecessor products can be used as a basis for identifying solution options (Buchert et al. 2016). This model-based approach shall now be demonstrated with the simple example of a pedelec frame.

Based on a previous LCA study, emissions for wrought material production were identified as an important lifecycle phase oriented property (Neugebauer et al. 2013). Hence, the indicator "CO_2 emissions in material production" was selected as a sustainability target for improving the pedelec frame. Furthermore, the durability of the pedelec frame in the use phase was chosen as a second target. The frame-durability determines a part of the value provided to the customer and may contribute to an overall reduction of CO_2 emissions if the lifetime of the pedelec is extended. Another reason for choosing durability as a target is to verify that a decision on material matters does not negatively affect the use-phase of the bicycle frame. Durability is a lifecycle-oriented property implying that influences (load and forces) from the respective lifecycle phase (use phase) are either assumed based on experience or on user studies or empirical studies of similar processes. The accurate determination of the frame durability would require a combination of different models for simulating the material strength under both static and dynamic load as well as for usage behaviour. In that pursuit, durability was examined by means of simplified analysis of axial frame deformation and v. Mises strength in comparison to tensile strength of the material. Figure 6 displays the causal relations between durability and CO_2 emission in material extraction with the product characteristics material and geometry.

Durability is dependent on the stiffness of the frame and on forces applied during use. The relations between material parameters such as young's modulus and stiffness follow principles of physics. The causal relations can thus be captured in

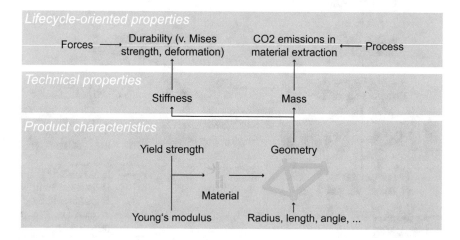

Fig. 6 Relations to calculate lifecycle oriented properties

mathematical equations. The v. Mises strength and the deformation can be calculated by an automatic FEM analysis. An existing frame CAD and FEM model from a predecessor product were utilised as a basis for the respective calculations. Since the new design may deviate from the original frame, the results calculated can only be understood as heuristic. Nevertheless, the process yields valuable insights about which materials are suitable for given requirements already in conceptual design, with the assumption that the frame design does not change significantly.

Figure 7 gives an example of how the data model for a decision support tool can be structured. The classes are instantiated for the selection of a material for the pedelec frame. The following dependencies between different classes of information were formalised:

- Constraints (broken down targets) are associated with product properties.
- Product properties can be calculated based on further properties, constants or characteristics.
- The CO_2 emission for wrought material, for example, can be calculated by the property *mass* times the constant *CO_2 emissions per kg* wrought material.
- The constant *CO_2 emissions per kg* can be derived from an environmental database, e.g. the ELCD database, thus through an IT-interface.
- IT-interfaces require certain data, in this case the ELCD material name, in order to yield the desired information *CO_2 emissions per kg* material.
- *Mass* is calculated by *volume* and *density* for the material.
- The *volume* can be easily calculated by a CAD system.
- Possible values for the characteristic (e.g. specific materials) are automatically derived from a repository which is linked to the model. In the case of the pedelec, all materials from the Siemens NX database were taken into consideration.

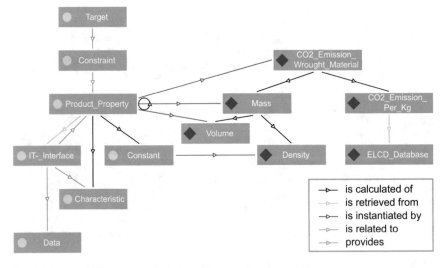

Fig. 7 Meta model for target calculation with exemplary instantiation

The software tool interprets all the interconnected steps independently, starting from a target and proceeding up until the point when it reaches an IT-database. The tool then derives the required information and successively inserts the information derived or calculated until a value for the product property concerning a selected constraint can then be calculated. The benefit of this approach is that all (discrete) values for a characteristic can be automatically iteratively assessed even if the required information is dispersed among different IT-tools. The IT-tools are accessed by respective interfaces e.g. a Service Oriented Architecture (SOA) interface. If all relevant relations are modelled according to the meta-model, the software tool can automatically calculate viable values for a characteristic and thus exclude unfitting solution options and reduce the solution space.

In the case of the pedelec frame, the tool proposed 5 different steel versions which fulfill the CO_2 emission and the deformation and strength constraints as a proof of concept. However, not all materials were listed in the ELCD database and were therefore excluded. Otherwise more suitable options could have been derived. The mapping of different material databases moreover remains imprecise and requires further research in order to boost accuracy. A more detailed description of this first model prototype can be found in the publication of Stark and Pförtner (2015).

A discussion of how the use case can be extended to assemblies and entire product systems can be found in Sect. 5.

4.3 Guidance for Achieving and Proving Compliance with Sustainability Targets

When sustainability design targets are formulated, engineers have to take action to prove that compliance with these targets in all stages of the PCP. Furthermore, guidance is necessary for assisting engineers in achieving the respective targets. These activities can be steered and supported by design methods. Since many methods for Sustainable Product Development (at least concerning environmental sustainability) are available, Ernzer and Birkhofer stated already in (2002) that: "the difficulty [...] is not the development of design methods anymore, but rather the ability to select the relevant methods." Hence, a scheme for selecting and combining methods was developed, which allocates a suitable approach to designing activities necessary for achieving or proving adherence with a sustainability target. The approach consists of a taxonomy of SPD methods and a method repository including 29 design methods. Figure 8 shows the three major steps for method selection and application. Step 1 characterises the definition of milestones. Milestones constitute a point of time in a design project where the achievement of a sustainability design target has to be proven.

A relevant sustainability target for a pedelec redesign process could be, for example, to decrease cost and CO_2 emissions in the usage phase with reference to what specific elements could be broken down into various alternatives for a drive

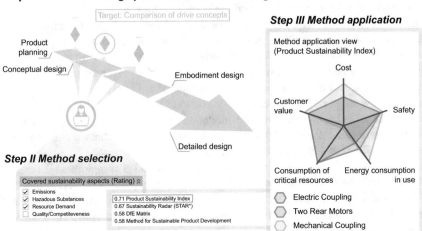

Fig. 8 Method selection and application feature for guiding engineers and to validate design performance against sustainability design targets

concept. Hence, a comparison of variants for the drive concepts regarding CO_2 emission and cost is necessary for the process of reporting results at a milestone towards the end of the conceptual design phase.

The second step (see Fig. 8) aims at selecting a suitable method for achieving targets defined in step 1. To that end, a taxonomy of design methods was put together (see Table 1).

The taxonomy outlines the design activity which the method supports (e.g. assessment), as well as information about the effort and information inputs required. Furthermore, it considers the type of targets which can be addressed by the method (e.g. addressed sustainability aspects or quantification of the target).

Complimentary to the taxonomy development, 29 SPD methods were successfully identified which were found to be compliant with predefined criteria (e.g. coverage of the whole product lifecycle, accessibility or a focus on technical products). Figure 9 shows an example for proving achievement of sustainability targets by selecting appropriate methods for the pedelec drive concept.

In addition to improved method selection, a concept for fostering application of methods was also implemented for three different methods (Pförtner et al. 2016). The main idea behind this approach was the development of an information platform which stores sustainability relevant information for a product and makes it available for the application of various methods. Only by following this approach does a combination of different SPD methods become attractive, since effort for information search can therein be reduced. Both the selection scheme and the information platform were implemented in the PDM system Siemens Teamcenter. Hence, necessary product-information (e.g. product structure, weight) can be imported. Further drawbacks and advantages of the approach are presented in Sect. 5.

Table 1 Taxonomy criteria for method selection

Criterion	Options
Method purpose	Identification of improvement measures, comparison, assessment, direct selection of product characteristics based on targets
Quantification	Qualitative, quantitative, semi-quantitative
Covered sustainability targets	Emissions, hazardous substances, resource demand, quality/competitiveness, safety, material origin, cost
User of the method	Product manager, product designer, sustainability expert
Effort for application	1 = low, 2 = middle, 3 = high
Necessary information for application	Requirements/functions, solution concepts, product architecture, CAD files/EBOM, production process/MBOM, auxiliary information
Redesign focus	Yes/no

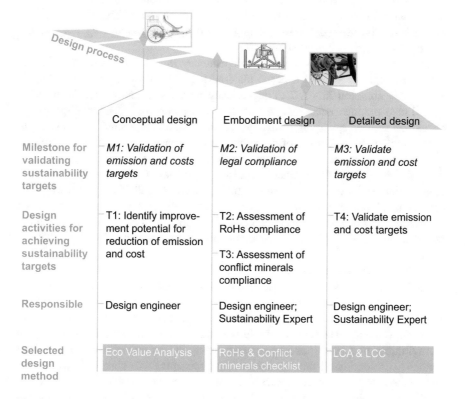

Fig. 9 Exemplary method selection results for a pedelec redesign process

5 Conclusion and Outlook

The last chapters presented different approaches on how the challenges for integrating sustainability targets into the design process (summarised in Fig. 3) can be addressed. For the specific case of modularisation, it was shown how the breaking down of sustainability design targets can be supported by qualitative causal diagrams (see Sect. 4.1). Nevertheless, qualitative visualisation of decision pathways can only be a first step towards decision support based on facts and data. What remains a challenge however, lies, in defining scenarios on how multiple sustainability design targets can be achieved by varying sub-targets for assemblies or subsystems. A lifecycle CO_2 reduction target could be, for example, achieved by material substitution of a pedelec frame or the more costly development of an efficient motor. To properly take stock of these side effects and trade-offs, a parametric model becomes necessary, which serves to establish connections between the decision criteria of the three sustainability dimensions. These "dependency models" can be utilised for setting targets but also for minimizing the solution spectrum of possible design solutions.

In Sect. 4.2, a first prototype of a dependency model was presented with the simple example of a pedelec frame material selection regarding technical targets and a CO_2 emission limits. The dependency model was represented in an ontological map and interpreted by a self-developed software tool. The model-prototype developed showed satisfying results, yet remains limited to a single component. In order to allow consideration of assemblies and complete products, more comprehensive models are necessary which comprise libraries of components from previous design projects that contain sustainability relevant information (e.g. material composition, GHG emissions, cost etc. (see Buchert et al. 2016). By following this approach, solution configurations can be identified which are compliant with a set of sustainability targets. In this context, model design must be simplified due to the fact that the effort for coupling different models in dispersed IT systems stands quite high.

In that pursuit of deeper understanding of the product's interrelation with sustainability impact, Sect. 4.3 presented a more process-oriented perspective on achieving sustainability design targets. By providing a selection scheme for SPD methods, the best suitable approach can be assigned to the tasks which are necessary for proving that sustainability design targets were achieved. In addition to the main findings of a corresponding literature analysis (see end of Sect. 2), a lack of methods considering all three sustainability dimensions was observed. While several methods focusing on environmental sustainability exist, approaches concerning social sustainability remain scarce. An integrated view of all three dimensions is indeed nearly non-existent. Furthermore, descriptions of several existing methods have only scratched the surface, while use cases for successful implementation are hard to come by. Nevertheless, the developed selection scheme and information platform presents the opportunity for combining heterogeneous approaches (such as qualitative guidelines and quantitative assessment methods) which allow for an overall more holistic perspective on the product.

References

Allenby, Braden R. 1991. Design for environment: A tool whose time has come. *SSA Journal* 12(9): 5–9.

Baumann, H., F. Boons, and A. Bragd. 2002. Mapping the green product development field: engineering, policy and business perspectives. *Journal of Cleaner Production* 10(5): 409–425. doi:10.1016/S0959-6526(02)00015-X.

Bretzke, Wolf-Rüdiger. 1980. Der Problembezug von Entscheidungsmodellen. 29: Mohr Siebeck.

Buchert, T., A. Kaluza, F.A. Halstenberg, K. Lindow, H. Hayka, and R. Stark. 2014. Enabling product development engineers to select and combine methods for sustainable design. *Procedia CIRP* 15: 413–418. doi:10.1016/j.procir.2014.06.025.

Buchert, T., A. Pförtner, J. Bonvoisin, K. Lindow, and R. Stark. 2016. Model-based sustainable product development. In *Proceedings of the 14th International Design Conference*, ed. Dorian Marjanovic, Mario Storga, and Neven Pavkovic, 145–154. Stanko Skec: Nenad Bojcetic.

Ciroth, Andreas, Jean-Pierre Theret, Mario Fliegner, Michael Srocka, Volker Bläsig, and Özlem Duyan. 2013. Integrating life cycle assessment tools and information with product life cycle management. In *Proceedings of the 11th Global Conference on Sustainable Manufacturing*, ed. Günther Seliger, 210–212. Universitätsverlag der TU Berlin.

Collado-Ruiz, Daniel, and Hesamedin Ostad-Ahmad-Ghorabi. 2013. Estimating environmental behavior without performing a life cycle assessment. *Journal of Industrial Ecology* 17(1): 31–42. doi:10.1111/j.1530-9290.2012.00510.x.

Duflou, Joost, and Wim Dewulf. 2005. Eco-impact anticipation by parametric screening of machine system components: An introduction to the EcoPaS methodology. In *Product engineering: Eco-design, technologies and green energy*, ed. Doru Talaba, and Thomas Roche, 17–30. Dordrecht: Springer Science + Business Media Inc.

Ernzer, M., and H. Birkhofer. 2002. Selecting methods for life cycle design based on the needs of a company. In *Proceedings of the 7th International Design Conference*, ed. Dorian Marjanovic, 145–154.

Gleisberg, Jochen, Oliver Knapp, Stefan Pötzl, and Michael Becker. 2012. Modular products—how to leverage modular product kits for growth and globalization. http://www.rolandberger.fr/publications/publications_monde/2012-04-10-Modular_Products.html.

Kara, Sami, Suphunnika Ibbotson, and Berman Kayis. 2014. Sustainable product development in practice: an international survey. *International Journal of Manufacturing Technology Management* 25(6): 848–872. doi:10.1108/JMTM-09-2012-0082.

Neugebauer, Sabrina, Julia Martinez-Blanco, René Scheumann, and Matthias Finkbeiner. 2015. Enhancing the practical implementation of life cycle sustainability assessment—proposal of a Tiered approach. *Journal of Cleaner Production* 102: 165–176. doi:10.1016/j.jclepro.2015.04.053.

Neugebauer, Sabrina, Ya-Ju Chang, Markus Maliszewski, Kai Lindow, Rainer Stark, and Matthias Finkbeiner. 2013. Life cycle sustainability assessment & sustainable product development: A case study on pedal electric cycles (Pedelec). In *Proceedings of the 14th Global Conference on Sustainable Manufacturing*, ed. Günther Seliger, 549–554. Universitätsverlag der TU Berlin.

Ny, Henrik, Sophie Hallstedt, Karl-Henrik Robèrt, and Göran Broman. 2008. Introducing templates for sustainable product development. *Journal of Industrial Ecology* 12(4): 600–623. doi:10.1111/j.1530-9290.2008.00061.x.

Pförtner, A., T. Buchert, K. Lindow, R. Stark, H. Hayka. 2016. Information management platform for the application of sustainable product development methods. *Procedia CIRP* 48: 437–442. doi:10.1016/j.procir.2016.04.091.

Shuaib, Mohannad, Daniel Seevers, Xiangxue Zhang, Fazleena Badurdeen, Keith E. Rouch, and I. S. Jawahir. 2014. Product sustainability index (ProdSI). *Journal of Industrial Ecology* 18(4): 491–507. doi:10.1111/jiec.12179.

Stark, Rainer, and Anne Pförtner. 2015. Integrating ontology into PLM-tools to improve sustainable product development. *CIRP Annals-Manufacturing Technology* 64(1): 157–160. doi:10.1016/j.cirp.2015.04.018.

Steffen, Will, Katherine Richardson, Johan Rockström, Sarah E. Cornell, Ingo Fetzer, Elena M. Bennett, Reinette Biggs, Stephen R. Carpenter, Wim de Vries, and Cynthia A. de Wit. 2015. Planetary boundaries: Guiding human development on a changing planet. *Science* 347(6223): 1259855. doi:10.1126/science.1259855.

Telenko, Cassandra, Julia M. O'Rourke, Carolyn C. Seepersad, and Michael E. Webber. 2016. A compilation of design for environment guidelines. *Journal of Mechanical Design* 138(3): 31102. doi:10.1115/1.4032095.

Weber, Christian, Horst Werner, and Till Deubel. 2003. A different view on product data management/product life-cycle management and its future potentials. *Journal of Engineering Design* 14(4): 447–464. doi:10.1080/09544820310001606876.

Johannes Seidel, Chair of Assembly Technology and Factory Management, Institute for Machine Tools and Factory Management, Technische Universität Berlin, Berlin, Germany

Value creation is understood as the comprehensive, activity-based framework within which transformation processes like manufacturing take place. Yet the results-oriented, economic definition of value creation is seen to be too narrow in the context of sustainable manufacturing. Value creation is therefore defined here as the tangible and intangible transformation processes in the pursuit of the creation of useful products and of the accumulation of intellectual capital given consideration of the sustainability criteria, indicators, and associated global living environments.

Global value creation gained significant momentum with the increased use of new communication and transportation techniques, a phenome-non often termed processes of globalisation. As a consequence of this transformation, manufacturing enterprises are confronted with increasing complexity coupled with a growing intensity of competition. Increased specialisation on core competencies as a consequence of this enhanced competition has made the division of value creation among a number of enterprises necessary. The organisation of networks that are spread all over the globe appears to be an inevitable step in the recent development of manufacturing practices. In this context, manufacturing activity can be seen as a value creation network (VCN) connecting value creation modules (VCM), each one defined not only by monetary or economic parameters but by social and environmental aspects as well. Sustainability is possible when every module of a VCN, or, at best the whole VCN as a system, is directed at increasing benefits for society and the environment while maintaining economic profitability. In this context, examining VCNs in their totality is therefore just as important as enhancing the sustainability of individual manufacturing technologies.

In the following chapters, such design approaches are described for or-ganisations and networks. In this pursuit, we aim at peering beyond the individual value creation module. The first chapter, *Future of business models in manufacturing*, explains the development of the sustainable business model (SBM)-concept within a globalised world and gives a preview of the manufacturing world of the future by combining SBM research with future studies. Product Service Systems and Circular Econ-omy-based business models are presented as examples that have the potential of meeting current and future sustainability challenges by applying a systems perspective on VCNs. The second chapter, *Industrial Symbiosis in value creation networks*, takes up the topic of Circular Economy with a specific look at how material is reused across industry and production lines. By applying the concept of Industrial Symbiosis, it presents a method that aims at closing material cycles not only within a company, but within a VCN and even across multiple VCNs that were originally independent from one other. The third chapter, *Integration of sustainability into the corporate strategy*, entails a concept for the restructuring of the entire organisation from tangible and intangible resources, business processes and the respective management disciplines. Finally, an integrated model-based framework is then presented that aims at enabling sustainability management and corporate sustainability performance measurement given the multidimensional requirements of VCNs and individual business fields.

Future of Business Models in Manufacturing

Johannes Seidel, Ana-Paula Barquet, Günther Seliger and Holger Kohl

Abstract In order to achieve systematic change in pursuit of sustainable manufacturing, both a strategic long-term perspective employing methods from future studies and a concrete implementation of the knowledge gained in sustainable business models are necessary. In this chapter, the concepts and exemplary methods for sustainable business model innovation are introduced with a special focus on sustainable manufacturing. Circular Economy-based business models and Product Service Systems are explained as examples of sustainable business models, along with a deduction of sustainability factors for both examples. The fruitful combination of future studies and sustainable business model development is illustrated in the example of a so-called *living factory*, a modular and adaptive production environment which integrates aspects of Circular Business Models and Product Service Systems.

Keywords Sustainable business models · Circular economy · Product service systems · Manufacturing scenarios

J. Seidel (✉) · A.-P. Barquet · H. Kohl
Institute for Machine Tools and Factory Management,
Technische Universität Berlin, Pascalstrasse 8-9, 10587 Berlin, Germany
e-mail: seidel@mf.tu-berlin.de

G. Seliger
Chair of Assembly Technology and Factory Management, Institute for Machine-tools
and Factory Management, Technische Universität Berlin, Berlin, Germany

© The Author(s) 2017
R. Stark et al. (eds.), *Sustainable Manufacturing*, Sustainable Production,
Life Cycle Engineering and Management, DOI 10.1007/978-3-319-48514-0_10

1   Introduction

Bringing the topics of business models, future studies and sustainability research together, this chapter puts itself in a relatively new tradition of manufacturing science. Since the 1990s, the literature in the three fields mentioned above has been growing indeed, yet opportune combinations of them so far remain rare. This being the aim of this text, short introductions into each field will be made, so that existing literature can be conveniently linked to our own contributions to research on sustainable business models and future studies.

Given the challenges which current modes of production and consumption place on nature and society, it seems necessary to pursue a new way of conducting business. Transforming business models into sustainable business models and creating pathways for sustainable technology development thus constitute the main themes of this chapter. In Sect 2, a short introduction of the inner-workings and benefits of sustainable business model concepts and tools will be given, before two specific examples, namely Product Service System-based and Circular Economy-based business models will be elaborated on. A special focus will lie on the analysis of sustainability factors for those two cases. Section 3 focuses on the tools for creating successful sustainable business models drawing on findings from the area of scenario planning as an instrument of future studies. This last chapter also presents the *Living Factory* as an exemplary result of combining future studies with business model innovation.

2   Sustainable Business Models

In the simplest terms, the concept of business models can be explained by splitting the term into its components. A business can be seen as the activity of buying and selling goods and services for the purpose of earning money, while a model is a means of representing reality in a structured, simplified and intelligible manner. A business model can ergo be understood as a structured, simplified and intelligible representation of how a company buys and sells goods or services and in that process, earns money. With this logic, a business model is a qualitative instrument for strategizing how business should be done. With the rise of the internet in the early 1990s, how business is being conducted has changed immensely. Value creation and communication networks have spread around the globe and diversified partners and consumer segments. At the same time, due to this development, both value creation and the predictability of a business's success has risen to a new level of complexity. Meanwhile, the first conceptualisations of how companies conduct their businesses have appeared and the term *business model* has arisen (Zott et al. 2011) as a means of describing how a business now operates. In the pursuit of assisting companies maintain competitive advantage by means of understanding, comparing, assessing, predicting and changing the way of doing business, diverse and even controversial concepts and approaches to business models have emerged in their wake. Mayo and Brown focus on the operational content, i.e. strategic purpose of a business by stressing the "key interdependent systems that create and sustain a competitive business" (Mayo and Brown 1999, 18). Morris, Schindehutte and Allen, on the other hand, propose a level-decision-approach by framing the supra-levels 'foundation,' 'proprietary,' and 'rules' levels á six sub-levels to lead business decision-making and to ensure that the individual decisions that are made within the company are internally consistent (Morris et al. 2005, 729). The three supra-levels cover the main areas of managerial decision-making in a company that answer increasingly specific questions at each level. At the foundation level, such basic questions have to be answered, as, how, for whom and by means of what sources of advantages, is value created? Furthermore, how exactly is profit generated? Meanwhile the proprietary level focuses on how the aspects of the foundation level are handled best and most uniquely. Finally, on the rules level, entrepreneurs should create guidelines and operating rules on how to strategize the foundation and proprietary of ones' business (Morris et al. 2005, 730f.). Osterwalder and Pigneur developed a value-based approach, in which the term business model entails a description of "the rationale of how an organization creates, delivers and captures value" (Osterwalder and Pigneur 2013, 14). This economic point of view allows an entrepreneur to develop and describe their business with nine core elements that involve this approach. These elements range from specific customer segments, revenues and partnerships to value proposition, activities and costs. Their business model approach is currently one of the most popular approaches for describing, developing and analysing business models.

Facing global environmental and social challenges, concepts like the business model of Osterwalder and Pigneur have been refined so that they include the reduction of negative impacts and the increase of benefits to both environment and society. Especially industries that thrive from non-renewable resources and those that create value mostly by employing cheap labour, serve as huge drivers of ecological imbalances and social inequalities. Concepts of sustainable business models are juxtaposed against the idea of 'business as usual' as they are meant to reflect upon their sustainability strategies and goals while earning money or replacing monetary earnings by environmental or social benefits in general. In that process, the meaning of value and the stakeholders involved in the business are redefined to be oriented towards the social and environmental perspective. In practice, that means that sustainability is not only implemented as a voluntary guideline, but as a fundamental part of each value proposition, value creation and value capture activity.

Product Service System-based and Circular Economy-based business models are examples of wide-ranging transformative models that include a product's entire lifespan into their considerations and are therefore viewed as the most effective sustainable business models. Their approaches require a perspective that is shifted from profit-oriented to enhanced benefits or reduced negative effects on environment and society.

2.1 Product Service System-Based Business Models: Satisfaction, Functionality and Ownership

The Product Service System (PSS) concept highlights the shift from traditional businesses based on the development and sale of physical products to a new business orientation based on functionalities and benefits delivered through a combination of products and services (Barquet 2015, 40f). Product Service Systems reflect on a long history of societal appreciation of service and ownership. After the world wars at the beginning of the 20th century, a noticeable development in the way people in the Western hemisphere organised their daily lives occurred which was interrelated with the changing socio-economic structures of that time. Domestic or commercial services like household servants or public laundry services were slowly replaced by self-service systems. In that process, a materialisation of services which is now fittingly represented by increasingly cheap goods like the washing machine, enabled households to complete housework at home without the help of external parties by buying a product instead of a service (Roy 2000, 291). Yet, all the while since the fifties, a convergence of product and service and a second reconfiguration of the product service-relation has taken place, which gives way to speculations about the renewed dematerialisation of the economic sphere and the emergence of a "new service economy in which profitability is based [...] on the provision of services to meet essential human needs" (Jackson 1996 quoted in Roy 2000, 292). Innovative combinations of products and services that can satisfy the same or even more needs than the product by itself, have appeared. In addition to car-sharing as a more prominent example of PSS, more unknown forms are beginning to enter the markets. Philips and Turntoo have, for example, created a PSS that sells light-per-lux and lightening systems with installation, maintaining and disposal, as an alternative to the ownership of lightening infrastructure, like cables and light bowls (Ellen MacArthur Foundation 2016). Those systems relieve the consumer of maintenance, insurance and disposal expenses while satisfying similar needs (in those cases transportation and light) as the original business model in which selling the product would have sufficed.

Tukker argued that beyond the rising numbers of researchers interested in this new set of PSS, such business models have attracted the attention of entrepreneurs once it became clear that characteristics and quality of a product were insufficient at holding onto a business's competitive advantage (Tukker 2015, 77). Designing and selling a combination of service and product now stands as a prominent value proposition. Manzini and Velozzi see "selling satisfaction instead of providing a product" (Manzini and Vezzoli 2003, 851) as the crucial element of PSS business models. Various benefits abound for companies, like reaching out to new market sectors (Allen Hu et al. 2012, 354). At the same time, consumers favour customised offers and the exemption from the responsibility for a product's end of life. In that vein then, PSS are not necessarily inherently sustainable, as there is no evidence that simply replacing product selling for service offer is sufficient for leading to more sustainable solutions (Evans et al. 2007, 4226). Of course, the lesser need for materials and resources during the manufacturing process on account of the higher span of consumers that can be reached with lesser products, the higher the efficiency employed. This might well therefore serve to reduce the negative effects on the environment. Yet this factor alone hardly suffices to qualify PSS as *sustainable*.

Following Tukker's classification of PSS-based business models, the conclusion can be drawn that the three main categories that are product-oriented services, use-oriented services and result-oriented services, all offer different opportunities but also include different limitations for the promotion of social and environmental well-being. Product-oriented PSS could optimise energy and resource consumption since service offers, e.g. as maintenance and repair, might increase the use phase of products. However, the traditional dynamic of selling as many products as possible and therefore causing negative environmental effects, remains firmly in place. Use-oriented PSS, which includes models of leasing, renting and pooling, might on the one hand lead to higher impacts due to less careful consumer behaviour, but on the other hand to extensive improvement of usage efficiency. The volume of impact reduction due to this efficiency increase varies between 30 and 50 %, in instances of car sharing, ski-renting, and laundry services and even up to 1000 % for drilling rental services. An even higher share of environmental benefits could be offered by result-oriented PSS, as this can be completely detached from product-oriented concepts. Examples could be payment-per-service unit-business models, like pay-per-copy copy shops or catering services, where a result is offered instead of a product. These models break the link between profit and production volume and reduce the incentive for large-scale production volumes and the accompanying resource consumption. Producing less to satisfy the needs of the same amount of consumers can significantly reduce the overall material usage. Nevertheless, using less materials, i.e. more durable materials, could be an incentive for result-oriented services (Tukker 2015, 86). To facilitate the identification of sustainable practices, a special set of five sustainability factors of PSS (see Fig. 1) was created. In combination, they target not only the environmentally thoughtful handling of resources, but also social justice and change.

(1) Design for Environment (DFE) is meant to include all stages of a product's lifecycle by following strategies of minimizing material and energy consumption and the selection of low impact materials and energy-efficient systems. What's more, cleaner technologies and environmentally friendlier materials and optimised distribution systems should be used.

Principles of disassembly, upgrading and adaptability should likewise be considered as end-of-life strategies. *(2) The identification of the value for each stakeholder* should take into account that longer lifespanmight decrease production, but cost savings can occur due to the reduction of material, the incentivizing of extended PSS lifecycles, and the profitability of new services. *(3) Promoting change in behaviour through educating consumers and PSS providers* can help to overcome the high symbolic value attached to owning a product and thereby increase the involvement of the consumers and employees as well as the satisfaction

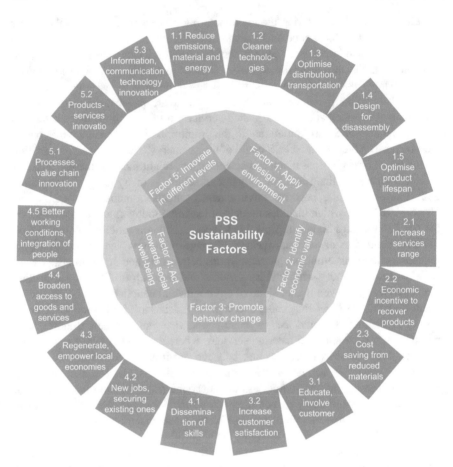

Fig. 1 Sustainability factors for product service systems

of the consumer's needs. Transparency, appearance, usability of the offer, price and time and cost saving can represent the means of this sustainability factor. As part of the *(4) Delineation actions to social well-being,* a PSS should also take responsibility for the creation and safety of jobs, for example, hiring and training employees to provide services. The fairness of the working-conditions (hours, wages, health and safety) and the tackling of social issues like the integration of social minorities or marginalised groups are also targets the attainment of sustainable PSSs. The empowerment of local communities and a broadened access to lower income segments should also constitute part of the actions for social well-being. *(5) Innovation in different levels* describes how innovations made in individual parts of the value chain might not be as sustainably successful as aligned and concentrated measures of innovation and optimisation. On-site assembly, remote controlling for maintenance and repair of products can be strategies for this factor (Barquet et al. 2016).

2.2   Zero-Waste or Reusable Waste: Circular Economic Business Models

Similar in their relevance and prominence in sustainable manufacturing are concepts of a circular economy that are based on the idea of following a product's whole life cycle and reducing resource input, waste, emission and energy leakage. Using nature as a model that cycles all its materials by means of natural decomposition and recreation, as promoted by *Industrial Ecology* thinkers like Keneth Boulding, Robert Ayres, Allen Kneese and Robert Frosch, involves putting money and hope into a product's durability and zero-waste policies.

Walter Stahel was one of the first scholars who, by introducing his concept of *Performance Ecology* in the 1980s, broached the issue of a closed-loop economy. Product-life prolonging measures like recycling, reusing, upgrading and remanufacturing coupled with a PSS-like idea of selling performance rather than the product, were to become the characteristics of his idea of a self-replenishing economy. William McDonough and Michael Braungart introduced their Cradle to Cradle (C2C) framework in the 1990s in Germany, wherein they argue that focusing on emission reduction is the wrong determination, as, emissions are the inevitable consequence of living. Instead, economy should focus on what they call materials-in-the-wrong-place-problems. Products should be designed and manufactured so that their materials could either be safely transformed in biological systems (*biological nutrients*), or be indefinitely recycled (*technical nutrients*), in case of substances that cannot be absorbed by nature. In the end, a cycled economy forms on account of the healthy waste that turns one process's waste into another process's resource. *The Blue Economy* as conceptualised by Gunther Pauli also stresses the importance of the question of how to create value from waste as a mean of providing for people's basic needs. The 2012 World Economic Forum shed new light on the idea of Circular Economy since the Ellen MacArthur Foundation introduced their publication *Towards the Circular Economy* and therein caused re-examination of previous ideas with a similar focus (Brennan et al. 2015, 223f).

A study of literature on circular business models (CBM) shows that they are generally considered to be sustainable. Five factors compounded out of 16 sub-factors seem to be critical for benefitting the environment and society while generating economic profit at the same time (see Fig. 2): *(1) Resource optimization* targets the saving of material, use of material and energy from renewable resources, dematerialisation, the creation of value from formerly considered waste and the creation of more value from each unit of resource (World Economic Forum 2014; Ellen MacArthur Foundation 2013a, b; Low et al. 2016; Geng et al. 2016; Schulte 2013; Winkler 2011; Guohui and Yunfeng 2012; Romero and Noran 2015). *(2) Improve environmental capabilities* consists of the reduction of negative emissions into the environment while increasing positive emissions to foster e.g. soil health and land productivity (World Economic Forum 2014; Ellen MacArthur Foundation 2013a, b). *(3) Risk reduction and control* can be achieved through design for

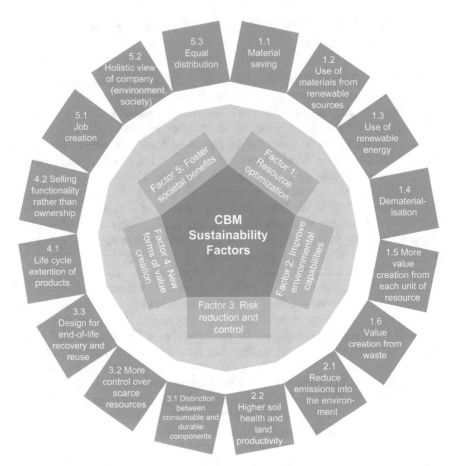

Fig. 2 Sustainability factors for circular business models

end-of-life recovery and reuse, whereby more control over scarce resources and a distinction between consumable and durable components can be attained (World Economic Forum 2014; Ellen MacArthur Foundation 2013a, b). *(4) New forms of value creation* can be reached by increasing the products' longevity, which then can foster new forms of consumption such as pay-per-use instead of ownership (Schulte 2013). *(5)* Finally, circular economic business models can *foster societal benefits* by creating new jobs, fostering equal distribution by fair wages and social thoughtful distribution of job opportunities, as well as by means of their holistic view of the company with regards to the environment and society (World Economic Forum 2014; Ellen MacArthur Foundation 2013b; Siemieniuch et al. 2015).

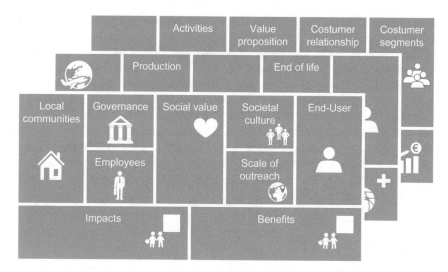

Fig. 3 Section of the three layered canvas business model tool (Joyce et al. 2015)

3   Developing Sustainable Business Models

Sustainable business model tools were developed to either adapt conventional business models or design new ones so that they fulfil the purpose of creating business that are environmentally and socially friendly as well as economically sufficient. Osterwalder and Pigneur developed the most common tool for business model design, called Canvas. In drawing up the nine core elements of their business model approach that was mentioned above (Costumer Segments, Channels, Costumer Relationship, Revenues, Value Proposition, Resources, Activities, Partners and Costs), entrepreneurs can easily conceptualise their business model (Osterwalder and Pigneur 2013). Criticism from environmentally and socially concerned academics and economists targets the focus on the economic perspective and benefits to the disadvantage of environmental and social issues. To meet this demand, the *three layered Canvas* (see Fig. 3) was later developed.

3.1   The (Three Layered) Canvas: A Tool for Sustainable Business Model Creation

Starting out with the idea that businesses will be more sustainable and also economically more successful when their business model innovations take a triple bottom line approach "people, planet and profit", as John Elkington imagined it in 1998, Joyce, Paquin and Pigneur designed a *triple layered canvas* that takes both economic, social and environmental benefits and impacts into account (Joyce et al. 2015).

The authors used elements of *Environmental Life Cycle Assessment* to create the environmental layer of their concepts, which now include Functional Value, Materials, Production, Supplies and Outsourcing, Distribution, Use Phase, End-of-Life, Environmental Impacts and Environmental Benefits. Using a *Stakeholder approach,* they designed the nine elements of their social layer (Social Value, Employees, Governance, Local Communities and Suppliers, Societal Culture, Scale of outreach, End-Users, Social Impacts and Social Benefits). Vertical coherence enables the comparison and analysis of interaction and interference of specific elements, like for example value proposition, functional value and social value (Joyce et al. 2015).

3.2 Business Model Innovation Meets Future Studies

The desire to know the future can be observed continuously throughout time. Independent of geographic or cultural boundaries, the practices range from highly spiritual (divination or prophecy) to purely scientific (probability calculation or game theory), and build hybrid forms like Utopian concepts in the arts or social sciences. After the Second World War, scientific future studies took a turn to what is now called modern future studies (Son 2015, 122f.). Scenario planning was introduced in the 1950s as a method of demonstrating the extremes and a variety of hypothetical futures, and in that pursuit, a shift from forecasting to the manageability of the outcome with present measures emerged (Son 2015, 124). Nowadays, scenario planning is used as a tool for describing possible future outcomes and situations based on a complex net of influence factors. A fragmentation of future studies brought a variation of approaches and goals, such as explorative or normative scenarios (Bradfield et al. 2005). Abele and Reinhart, for example, created explorative scenarios for the German manufacturing industry in 2020 and described possible futures surrounding fields in which a high level of adaptability and competitiveness with regards to the global markets is required (Abele and Reinhart 2011). Using the pathways for sustainable technology development approach by Gausemeier, their findings were used to deduce the concept of a highly modern and versatile factory based on modular machine tools, the so-called "Living Factory" (see Fig. 4) (Gausemeier 2014). A Living Factory involves high versatility and mobility of production facilities that can be reached through the combination of modular machine tool frames, so-called LEG^2O frames, and business model innovation that makes use of a Product Service System and circular business model concepts. A detailed description and analysis of the LEG^2O frame is presented in the part *Sustainability-driven development of manufacturing technologies* in this book. Lightweight constructed and accuracy-tuned modular machine tools enable partial replacement and flexible combination. Applying a PSS-based system might mean renting or leasing the machine-modules, which are in the best case provided

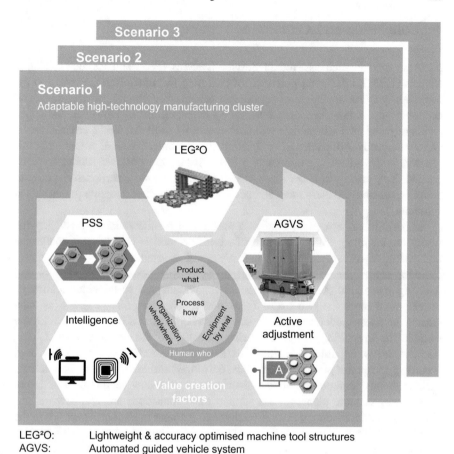

LEG²O: Lightweight & accuracy optimised machine tool structures
AGVS: Automated guided vehicle system

Fig. 4 Excerpt of an abstract representation of the living factory

according to the principles of a circular economy, along the lines of occupancy and requirement. A Living Factory can therefore adapt itself to fluctuations in demand and environmental and social conditions. Intelligent communication and information technology is used, including RFID tags and automated guided vehicles for logistics. Specifically, this means to reach a circular system in which machine-modules are offered by means of a central technology provider who can assist in building up the initial modular machines, and later on extend on them by adding additional building blocks, or updating them with new functionalities and smart blocks. Similarly, unused building blocks can then be taken back to be transferred to another customer. Outdated building blocks, meanwhile, can be updated, remanufactured or recycled for material recovery by the central technology provider.

4   Conclusion and Outlook

Business models such as Product Service Systems (PSS) and Circular Business Models (CBM), offer great potential for changing manufacturing according to the triple-bottom line approach of producing benefits for society, environment and economy and at the same time minimizing negative effects. However, the application of those business models will not necessarily fulfil economic, environmental and social needs. Adherence to such factors like the ones that were presented in this chapter, is nevertheless essential if a truly sustainable business model is to be created. Yet, getting to know these factors might stimulate enterprises not only to adopt sustainable business models, but also to implement sustainable practices and solutions.

Scenario planning can be seen as a useful tool for theoretically predicting the future's needs along with the success of a business model. The complex challenges that businesses and sustainability will face are well advised to be included in current business model innovation in pursuit of enhancing sustainability success and reducing risk of failure. Business model innovation and sustainable technology development mark the two major fields that require scientific progress, as, sustainable business models indeed rely heavily on both aspects. Both also include new ideas in the structuring of manufacturing processes as the example of the Living Factory shows. Modular machine-tools that are themselves produced and used according to circular principles need to be developed and tested. The transition from traditional business models to sustainable ones and how methods from future studies, e.g. scenario planning, can support these transitions are, furthermore, relevant subjects demanding deeper investigation. Another important aspect lies in the creation of indicators to measure the sustainability of business models. Building on the predominantly qualitative factors of developing quantitative approaches, has yet to be explored. The adoption of PSS and circular economy principles, moreover, can facilitate yet hardly guarantee that this version of business practice will result in a more sustainable performance. The need for future research likewise extends to the management of remanufacturing and (re-)consumption, which specifically requires a more transdisciplinary approach.

References

Abele, E., and G. Reinhart. 2011. Zukunft der Produktion: herausforderungen, forschungsfelder, chancen. München: Carl Hanser Verlag.

Allen Hu, H., S.H. Chen, C.W. Hsu, C. Wang, and C.L. Wu. 2012. Development of sustainability evaluation model for implementing product service systems. *International Journal Environment Sciences Technology* 9(2): 343–354. doi:10.1007/s13762-012-0037-7.

Barquet, Ana P. 2015. Creation of product-service systems (PSS) proposals in the fuzzy front-end. PhD dissertation, University of São Paulo, Brazil.

Barquet, Ana P., Johannes Seidel, Günther Seliger, and Holger Kohl. 2016. Sustainability factors for PSS business models. *Procedia CIRP* 47: 436–441. doi:10.1016/j.procir.2016.03.021.

Bradfield, Ron, George Wright, George Burt, George Cairns, and Kees van der Heijden. 2005. The origins and evolution of scenario techniques in long range business planning. *Futures* 37(8): 795–812. doi:10.1016/j.futures.2005.01.003.

Brennan, Geraldine, Mike Tennant, and Fenna Blomsma. 2015. Business and production solutions: Closing loops and the circular economy. In *Sustainability. Key Issues*, 219–239.

Ellen MacArthur Foundation. 2013a. *Towards the circular economy: Economic and business rationale for an accelerated transition*.

Ellen MacArthur Foundation. 2013b. *Towards the circular economy: Opportunities for the consumer goods sector*.

Ellen MacArthur Foundation. 2016. Selling Light as a Service. https://www.ellenmacarthurfoundation.org/case-studies/selling-light-as-a-service.Accessed 02 Aug 2016.

Evans, Stephen, J. Paulo Partidário, and Joanna Lambert. 2007. Industrialization as a key element of sustainable product-service solutions. *International Journal of Production Research* 45(18–19): 4225–4246. doi:10.1080/00207540701449999.

Gausemeier, P. 2014. Nachhaltige Technologiepfade für unterschiedliche Entwicklungsniveaus. PhD dissertation, Technische Universität Berlin.

Geng, Yong, Tsuyoshi Fujita, Hung-suck Park, Anthony S. Chiu, and Donald Huisingh. 2016. Recent progress on innovative eco-industrial development. *Journal of Cleaner Production* 114: 1–10. doi:10.1016/j.jclepro.2015.09.051.

Guohui, Song, and Li Yunfeng. 2012. The effect of reinforcing the concept of circular economy in west china environmental protection and economic development. *Procedia Environmental Sciences* 12: 785–792. doi:10.1016/j.proenv.2012.01.349.

Joyce, Alexandre, Raymond Paquin, and Yves Pigneur. 2015. The triple layered business model canvas. A tool to design more sustainable business models. *In ARTEM organizational creativity international conference, Nancy, France*.

Low, Jonathan S.C., Tobias B. Tjandra, Wen F. Lu, and Hui M. Lee. 2016. Adaptation of the product structure-based integrated life cycle analysis (psila) technique for carbon footprint modelling and analysis of closed-loop production systems. *Journal of Cleaner Production* 120: 105–123. doi:10.1016/j.jclepro.2015.09.095.

Manzini, E., and C. Vezzoli. 2003. A strategic design approach to develop sustainable product service systems: Examples taken from the 'environmentally friendly innovation' Italian prize. *Journal of Cleaner Production* 11(8): 851–857. doi:10.1016/S0959-6526(02)00153-1.

Mayo, Michael, and Gordon Brown. 1999. Building a competitive business model. *Ivey Business Journal* 63(3): 18–23.

Morris, Michael, Minet Schindehutte, and Jeffrey Allen. 2005. The entrepreneur's business model: Toward a unified perspective. *Journal of Business Research* 58(6): 726–735. doi:10.1016/j.jbusres.2003.11.001.

Osterwalder, Alexander, and Yves Pigneur. 2013. *Business model generation: A handbook for visionaries, game changers, and challengers*, 1st ed. Hoboken, NJ: Wiley.

Romero, David, and Ovidiu Noran. 2015. Green virtual enterprises and their breeding environments. Engineering their sustainability as systems of systems for the circular economy. *IFAC-PapersOnLine* 48(3): 2258–2265. doi:10.1016/j.ifacol.2015.06.424.

Roy, Robin. 2000. Sustainable product-service systems. *Futures* 32: 289–299.

Schulte, Uwe G. 2013. New business models for a radical change in ressource efficiency. *Environmental Innovation and Societal Transitions* 9: 43–47. doi:10.1016/j.eist.2013.09.006.

Siemieniuch, C.E., M.A. Sinclair, and M.J.d. Henshaw. 2015. Global drivers, sustainable manufacturing and systems ergonomics. *Applied Ergonomics* 51: 104–19. doi:10.1016/j.apergo.2015.04.018.

Son, Hyeonju. 2015. The history of Western futures studies: An exploration of the intellectual traditions and three-phase periodization. *Futures* 66: 120–137. doi:10.1016/j.futures.2014.12.013.

Tukker, Arnold. 2015. Product services for a resource-efficient and circular economy—a review. *Journal of Cleaner Production* 97: 76–91. doi:10.1016/j.jclepro.2013.11.049.

Winkler, H. 2011. Closed-loop production systems—a sustainable supply chain approach. *CIRP Journal of Manufacturing Science and Technology* 4(3): 243–246. doi:10.1016/j.cirpj.2011.05.001.

World Economic Forum. 2014. *Towards the circular economy: Accelerating the scale-up across global supply chains*. Geneva.

Zott, Christoph, Raphael Amit, and Lorenzo Massa. 2011. The business model: Recent developments and future research. *In Journal of Management* 37: 1019–1042.

Material Reutilization Cycles Across Industries and Production Lines

Friedrich A. Halstenberg, Jón G. Steingrímsson and Rainer Stark

Abstract The concept of Industrial Symbiosis aims at organizing industrial activity like a living ecosystem where the by-product outputs of one process are used as valuable raw material input for another process. A significant method for the systematic planning of Industrial Symbiosis is found in input–output matching, which is aimed at collecting material input and output data from companies, and using the results to establish links across industries. The collection and classification of data is crucial to the development of synergies in Industrial Symbiosis. Public and private institutions involved in the planning and development of Industrial Symbiosis rely however on manual interpretation of information in the course of creating synergies. Yet, the evaluation and analysis of these data sources on Industrial Symbiosis topics is a tall order. Within this chapter a method is presented which describes value creation activities according to the Value Creation Module (VCM). They are assessed before they are integrated in Value Creation Networks (VCNs), where alternative uses for by-products are proposed by means of iterative input-output matching of selected value creation factors.

Keywords Circular economy · Industrial symbiosis · Industrial ecology · Value creation networks · Input-output matching

1 Closing Material Cycles in Manufacturing

Industrial sustainability is a topic which can be addressed from a range of angles, including not only from the usual product and process perspective, but also on the level of Value Creation Networks (VCNs).

F.A. Halstenberg (✉) · J.G. Steingrímsson
Institute for Machine Tools and Factory Management,
Technische Universität Berlin, Berlin, Germany
e-mail: f.halstenberg@tu-berlin.de

R. Stark
Chair of Industrial Information Technology, Institute for Machine-tools and Factory Management, Technische Universität Berlin, Berlin, Germany

© The Author(s) 2017
R. Stark et al. (eds.), *Sustainable Manufacturing*, Sustainable Production,
Life Cycle Engineering and Management, DOI 10.1007/978-3-319-48514-0_11

The concept of a *Circular Economy*, which seeks to decouple global economic development from finite resource consumption, has attracted a lot of attention in recent years (Ellen MacArthur Foundation 2015). *Circular Economy* is an umbrella term for different material recovery techniques such as reusing, remanufacturing, and recycling (Ellen MacArthur Foundation 2015), as well as frameworks for closed material systems, such as the *Blue Economy* (Pauli 2010), *Industrial Ecology* (IE) (Frosch and Gallopoulos 1989), and *Industrial Symbiosis* (Chertow 2007). Within these frameworks, one can distinguish whether the material has been recovered from an intentionally manufactured product at the point of its end-of-life (EOL), or as a by-product (an unintended derivative of the production process). The waste framework directive of the European Commission specifies the hierarchy of waste from the least favourable option to the most favourable option (landfilling, energy recovery, recycling, reuse/remanufacturing, minimization, and prevention) (European Commission 2008). Since the term 'waste' conveys no or little value, the authors opt for the term 'by-product,' with its reference to originally unintended derivatives of manufacturing resulting separately from the desired product through industrial processes.

The term Industrial Ecology (IE) was coined by Frosch and Gallopoulos to depict the design of manufacturing entities analogous to natural ecosystems (Frosch 1992). As a sub-discipline of IE, Industrial Symbiosis is concerned with resource optimization among collocated companies (Jacobsen 2006). Industrial Symbiosis brings together traditionally separate industries into a collective approach for competitive advantage involving physical exchange of materials, energy, water, and/or by-products (Chertow 2000). In other words, Industrial Symbiosis aims at organizing industrial activity like that of a living ecosystem, where the by-product outputs of one process are used as valuable raw material input for another process. In an ideal Industrial Symbiosis, waste material (by-products) and energy are shared or exchanged between the actors of the system, therein reducing the net consumption of raw material and energy inputs, and thus the generation of waste and emissions (Sokka 2011). The geographic co-location of production plants with possible synergies in terms of waste streams, furthermore, serves to facilitate the exchange of the physical flows that are involved (Duflou et al. 2012). One aspect of the Factory of the Future, described by Herrmann et al. entails the symbiotic integration of factories into their surroundings (Herrmann et al. 2014). Cerdas et al. introduce the concept of a Circulation Factory, combining manufacturing with remanufacturing and recycling into an integrated system (Cerdas et al. 2015).

The term 'eco-industrial park' (EIP) describes, in a general sense, an industrial or a commercial area that is used by different companies. EIPs are networks comprising a variety of firms with an immediate geographical proximity to one another, where material exchange is carried out. An important precondition for an EIP is mutual trust, which seems to be a precondition to implementing common exchange relationships successfully (Bauer 2008; Hauff et al. 2012; Ludwig 2012). The EIP in Kalundborg, Denmark, is considered to be a seminal example in the literature on Industrial Symbiosis. The development of Industrial Symbiosis has been described as an evolutionary process in which a number of independent by-product exchanges have gradually evolved into a complex web of symbiotic interactions between five collocated companies and the local municipality (Ehrenfeld and Gertler 1997).

Results have shown that significant environmental savings are related to Industrial Symbiosis in Kalundborg (Jacobsen 2006). For example, three million m^3 of water could be saved through recycling and reuse. The environmental benefits of Industrial Symbiosis have been quantified in numerous further cases (Kincaid and Overcash 2001; Chertow and Lombardi 2005). Although Industrial Symbiosis has developed into a notable research topic, its impact on actual industrial practice remains very modest (Chertow 2007). Efforts by public and private institutions have been made to improve the systematic planning and development of Industrial Symbiosis over the past decades (Lowe 2007). Practitioners moreover consider it crucial to finding ways of obtaining buy-ins from businesses—an essential step for success. Many practitioners have noted the significance of company champions (Chertow and Park 2016) as well as the importance of using the language of business (costs, revenues, risk, etc.) to generate this buy-in (Laybourn 2015). Duflou et al. argue that 'the most effective way of strengthening Industrial Symbiosis is to increase the economic motivation' (Duflou et al. 2012).

A significant method for the systematic planning of Industrial Symbiosis is input–output matching. It is aimed at collecting material input and output data of companies, and using the results to establish links across industries. As an outcome of the method, a resource input associated with one organization can be matched to a complementary resource output of another organization (Lowe 2007). In the case of a certain proximity of a match, an integrated input–output matching method can also be recommended for a further conversion or treatment process (Bin et al. 2015).

Regarding the support of input–output matching, a growing trend has surfaced, whereby the application of internet-based IT tools such as Synergie by International Synergies, or the Resource-eXchange-Platform as part of the ZeroWIN EU project have emerged to further promote coordination and exchanges. Additional tools include Knowledge-Based Decision Support System (Boyle and Baetz 1998), Dynamic Industrial Materials Exchange Tool (Shropshire et al. 2000), Match Maker! (Chertow 1997), Industrial Ecology Planning Tool (Nobel and Allen 2000), WasteX (Clayton et al. 2002), Industrial Ecosystem Development Project (Kincaid and Overcash 2001), Residual Utilization Expert System (Fonseca et al. 2005), Institute of Eco-Industrial Analysis Waste Manager (Sterr and Ott 2004), Industrie et Synergies Inter-sectorielles (Massard and Erkmann 2007), SymbioGIS (Massard and Erkmann 2009), and Core Resource for Industrial Symbiosis Practicioners (Laybourn and Morrissey 2009).

The collection and classification of data is crucial to the development of synergies in Industrial Symbiosis (Cecelja 2016). Public and private institutions involved in the planning and development of Industrial Symbiosis rely on manual interpretation of information in the course of personal communication and case-by-case analysis. Cecelja et al. (2014) report that in the course of their service offer, practitioners access and interpret data collected from the industry by combining it with further data stored in databases such as the following:

- Proprietary databases built to monitor the activity of industry, e.g. industrial sectors, industrial volumes, planning and marketing datasets, and occasional project management technologies, such as environmental records, quality management practices, or
- Custom-made databases that offer access to case studies, e.g. Crisp system (Grant et al. 2010).

Bin et al. propose a big data analytics approach for developing industrial symbioses in large cities. The authors suggest that data can be acquired from structured or unstructured sources. Structured sources include company registration, waste exchange registry databases, the national pollutant emissions inventory, geographical information systems (e.g. Google Maps), lifecycle inventory databases, etc. Examples of unstructured data sources are financial reports, information from company websites, online news, social media, online encyclopaedias, and journal corpus (Bin et al. 2015).

The evaluation and analysis of these data sources regarding Industrial Symbiosis is of course challenging to say the least. Firstly, data has to be interpreted in the context of specific knowledge domains. Secondly, the resulting implications have to be evaluated in combination with available data about the surrounding value creation network (e.g. materials, technologies and objectives, environmental effects, economic and social benefits). Given increasing numbers of network participants, their dynamic behaviour within the network (e.g. inclusion of new technologies, inclusion of additional stages for by-product pre-processing, pre-treatment, transportation, and storage) and the resulting complexity of material streams, it becomes quite apparent that a systematic and thorough analysis through manual manipulation of data is outright impossible (Desrochers 2004; Mirata and Emtairah 2005). Furthermore, Grant et al. criticize the available datasets as outdated and incapable of assisting innovation (Grant et al. 2010).

In order to involve businesses in Industrial Symbiosis, online platforms for facilitating exchange of by-products have been provided. Industry organizations such as the United States Business Council for Sustainable Development (USBCSD), or facilitators such as National Industrial Symbiosis Programme (NISP), allow businesses a secure and common platform for discussing potential synergies through symbiosis (Chertow and Park 2016). In recent approaches, novel concepts such as ontology engineering have been introduced in matching tools and platforms for Industrial Symbiosis, since they can help to put tacit knowledge out there—essential for the mutual, nonmarket interactions required for Industrial Symbiosis (Cecelja et al. 2014; Cecelja 2016). Halstenberg et al. suggest employing organisational data systems such as Product Data Management Systems (PDM), Product Lifecycle Management (PLM), Enterprise Resource Management Systems (ERP). Utilizing these data for Input-Output Matching tools and platforms can add functionality to existing approaches (Halstenberg et al. 2016).

2 Method Design for Sustainable Manufacturing by Analysis of Value Creation Factors

A number of different approaches exist which address the issue of Match-Making for Industrial Symbiosis. In this section, the method for designing in pursuit of resource efficient approaches stemming from the domain of sustainable manufacturing is presented, involving analysis of value creation factors. The method relies on the concept of the Value Creation Module (VCM), which will be explained in Sect. 2.1, followed by a description of the method (Sect. 2.2).

2.1 The Value Creation Module (VCM)

Any type of value creation activity can be characterised in terms of a so-called value creation module (VCM) (Seliger 2008). The VCM is depicted in Fig. 1. A VCM is composed by five Value Creation Factors (VCF): product, process, equipment, organisation and human. Networks and modules are conceivable at different levels of aggregation (Wiendahl et al. 2009) (e.g. grinding a turbine blade, assembling a turbine, building a power plant, and providing power for a community), each with sustainability indicators that are identical on all aggregation levels or relevant for the respective aggregation level. Effective and efficient VCFs must be identified, combined into promising VCMs and promoted.

Fig. 1 Value Creation Module (VCM) (Seliger 2008)

A **Product** represents a desired, manufactured output according to design requirements, specifications and standards (Laperrière and Reinhart 2014). A **process** is understood as a task that depicts how desired outputs are created from inputs. **Equipment** is the means to manufacture the products, e.g. machine tools, jigs and fixtures, tools and measuring equipment. The crucial precondition for factory operations are **humans**. They are the direct employees involved in value creation, using qualifications and training to that end (Westkämper 2006). The **organisation** represents the functional, spatial and temporal context in which manufacturing tasks are carried out and managed (Spur 1994).

2.2 Description of the Method

This sub-section highlights the procedure for the method of designing for sustainable manufacturing and thereby included resource efficiency by means of analysis of value creation factors. A flowchart of the procedure can be seen in Fig. 2. The goal of the method is to model and plan value creation networks in a sustainable manner with a specific focus on by-product exchanges in the sense of Industrial Symbiosis objectives.

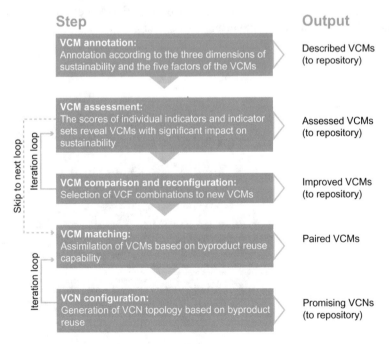

Fig. 2 Flowchart of the method for designing for sustainable manufacturing by means of analysis of value creation factors

Firstly, VCMs are annotated, assessed and improved. This part of the method focused for that reason, on the specific processes and not on their network. Secondly, the method focuses on the network level. Here the individual VCMs are matched in order to form Value Creation Networks (VCNs).

The VCM (see also Sect. 2.1) provides a structured framework for the **annotation** of value creation activities in the first step of the method. It allows the integrating of various levels of aggregation from a single manufacturing tool and operations via manufacturing cells and systems, whole factories with national and international entrepreneurial conglomerates or knowledge generating communities (Wiendahl et al. 2009). As this method prescribes, VCMs are annotated according to the three pillars of sustainability as well as according to the five VCFs (product, process, equipment, human and organisation).

In order to gain general knowledge on the sustainability performance of the VCM, a **VCM assessment** is performed in the second step. The scores of individual indicators and indicator sets reveal which of the VCMs have a significant impact on sustainability. In order to reduce dependency on detailed performance data, a qualitative approach is used. This approach enables a rapid cross-industry assessment of VCMs, capable of showing concrete improvement potential. The VCM assessment is based on the Bellagio principles and is requisite for a dynamic shift between module and network perspective.

The third step, **VCM comparison and reconfiguration** of alternatives, is then performed in order to eliminate shortcomings of VCMs which have been identified through the VCM assessment. In this step of the method, alternative comparison and VCM reconfiguration are conducted. Next, alternative comparison is performed by comparing the VCM assessment scores for two or more different VCMs. All VCMs are then described according to a VCM annotation structure, where elements and elements instances are utilised. This is made possible through similarity matching between these elements and elements instances. Depending on the similarity score of the selected comparison elements, the VCM can be reconfigured and its sustainability performance enhanced. The comparison criteria are selectable based on the VCM annotation and a reference VCM.

Once the comparison criteria and the similarity matching threshold have been determined, the highest scoring VCMs are presented, based on the individual indicator sets. The indicator set score is based on the VCM assessment. The VCM reconfiguration is a process for improving a reference VCM by VCF substitution. VCF of higher scoring VCMs are used for the process.

Figure 3 presents a comparison between two VCMs, 'Bamboo frame manufacturing at PTZ' and 'SUW sharing platform'. The latter offers significant improvements in public reach, which when implemented as a 'Help for self-help bamboo frame manufacturing in Vietnam' presents an improved overall sustainability performance.

The method focuses on the network implementation of the previously annotated, assessed and reconfigured VCMs. All VCMs considered are now treated as black-boxes, and matched with the purpose of forming networks. A network can be formed and planned according to various goals. The method presented focuses on

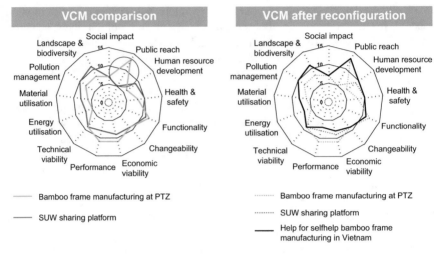

Fig. 3 VCM comparison and reconfiguration

the aspect of creating symbiotic relationships among companies in the sense of an Industrial Symbiosis.

The process of **VCM matching** begins with a classification of the by-product, all the while ensuring representation in a manner that is appropriate to the various industries. For example, a by-product can be classified as a biomaterial or a technical material (metals, ceramics, organic polymers, composites, semi-conductors and advanced materials). In the next step, the by-product is annotated in terms of quantitative and qualitative information. In this process, the VCM is described in a more detailed manner through information embedded in the VCM ontology belonging to the VCF taxonomies for product, process and equipment. The goal is an annotation which ensures that a by-product of one manufacturing entity is described in a suitable manner so that it can find a suitable fit with another manufacturing entity. The material type classification, economic factors, environmental considerations and known reutilisation possibilities are all required (e.g. stream behaviour, material cost, level of toxicity, reutilization possibilities) in that pursuit.

A **match of one VCM to another** is performed by comparing the respective in- and outputs. In order to establish possible usage, the by-product material stream is classified.

Moreover, in pursuit of identifying suitable relationships between VCMs within the considered VCN, an **input-output matching** approach is carried out to pair VCMs based on their by-products. For this purpose, a similarity algorithm is utilized. An important aspect of input-output matching is the range of matching since, depending on the type of description, different ranges are possible. In the case of a quantitative description, the pairing up can either be a 1-to-1 match or be located within a certain range. In the case of a qualitative matching, the inputs and outputs can be matched according to semantic descriptions.

According to the VCM matches identified, suitable **VCNs** have to be **configured** in the next step. From a single VCM, pairs of VCMs are generated and a network is formed by moving with the flow of by-products. Having the role of a broker in place—that is, a neutral network administrator who has the responsibility of creating a VCN and identifying open interfaces for new VCMs creation—is seen as a useful function for the arrangement of the different VCMs in a network. Three tasks are then performed in order to establish the networks. First, a joint effort opportunity is to be detected and promoted by a broker, made through an online platform. Then the main features best suited to describing the joint effort are to be classified. Finally, rough planning for the network is to be conducted. Possible network partners and their ideal locations can then be identified. A VCN topology is created by selecting one VCM to act as an anchoring point and other VCMs arranged accordingly.

References

Bauer, J. 2008. Industrielle Ökologie: Theoretische Annäherung an ein Konzept nachhaltiger Produktionsweisen. Dissertation, University of Stuttgart.

Bin, Song, Yeo Zhiquan, Low Sze Jonathan, Derek K. Choong, Denis Kurle Jiewei, Felipe Cerdas, and Christoph Herrmann. 2015. A big data analytics approach to develop industrial symbioses in large cities. *Procedia CIRP* 29: 450–455.

Boyle, C.A., and B.W. Baetz. 1998. A prototype knowledge-based decision support system for industrial waste management: Part I. The decision support system. *Waste Management* 18(2): 87–97.

Cecelja, F. 2016. Ontology evaluation for reuse in the domain of process systems engineering. *Comuputers & Chemical Engineering* 85: 177–187. doi:10.1016/j.compchemeng.2015.12.003.

Cecelja, F., T. Raafat, N. Trokanas, S. Innes, M. Smith, A. Yang, Y. Zorgios, A. Korkofygas, and A. Kokossis. 2014. e-Symbiosis: Technology-enabled support for industrial symbiosis targeting small and medium enterprises and innovation. *Journal of Cleaner Production*. doi:10.1016/j.jclepro.2014.08.051.

Cerdas, Felipe, Denis Kurle, Stefan Andrew, Sebastian Thiede, Christoph Herrmann, Yeo Zhiquan, Low Sze Jonathan, Song Bin Choong, and Sami Kara. 2015. Defining circulation factories—a pathway towards factories of the future. *Procedia CIRP* 29: 627–632.

Chertow, Marian R. 1997. The source of value: An executive briefing and sourcebook on industrial ecology. *Journal of Industrial Ecology* 2: 151–152. doi:10.1162/jiec.1997.1.2.151.

Chertow, Marian R. 2000. Industrial symbiosis: Literature and taxonomy. *Annual Review of Energy and the Environment* 25(1): 313–337. doi:10.1146/annurev.energy.25.1.313.

Chertow, Marian R. 2007. "Uncovering" industrial symbiosis. *Journal of Industrial Ecology* 11 (1): 11–30. doi:10.1162/jiec.2007.1110.

Chertow, Marian R., and D.R. Lombardi. 2005. Quantifying economic and environmental benefits of co-located firms. *Environmental Science and Technology* 39(17): 6535–6541. doi:10.1021/es050050.

Chertow, Marian R., and Jooyoung Park. 2016. *Scholarship and practice in industrial symbiosis: 1989–2014:* Springer Gabler.

Clayton, Anthony, John Muirhead, and Han Reichgelt. 2002. Enabling industrial symbiosis through a web-based waste exchange. *Greener Management International* 2002(40): 93–106.

Desrochers, Pierre. 2004. Industrial symbiosis: The case for market coordination. *Journal of Cleaner Production* 12(8): 1099–1110.

Directive 2008/98/EC of the European Parliament and of the Council of 19 November 2008 on waste and repealing certain Directives. European Commission. 312. Official Journal of the European Union L.

Duflou, Joost R., John W. Sutherland, David Dornfeld, Christoph Herrmann, Jack Jeswiet, Sami Kara, Michael Hauschild, and Karel Kellens. 2012. Towards energy and resource efficient manufacturing: A processes and systems approach. *5CIRP6 Annals—Manufacturing Technology* 61(2): 587–609. doi:10.1016/j.cirp.2012.05.002.

Ehrenfeld, John, and Nicholas Gertler. 1997. Industrial ecology in practice: The evolution of interdependence at Kalundborg. *Journal of Industrial Ecology* 1(1): 67–79.

Ellen MacArthur Foundation. 2015. Towards a circular economy: Business rationale for an accelerated transition. http://www.ellenmacarthurfoundation.org/publications/towards-a-circular-economy-business-rationale-for-an-accelerated-transition.

Fonseca, Daniel J., Eric Richards, Derek Williamson, and Gary P. Moynihan. 2005. A knowledge-based system for the recycling of non-hazardous industrial residuals in civil engineering applications. *Expert Systems* 22(1): 1–11.

Frosch, Robert A. 1992. Industrial ecology: A philosophical introduction. *Proceedings of the National Academy of Sciences* 89: 800–803.

Frosch, Robert A., and Nicholas E. Gallopoulos. 1989. Strategies for manufacturing. *Scientific American* 261(3): 144–152.

Grant, Gabriel B., Thomas P. Seager, Guillaume Massard, and Loring Nies. 2010. Information and communication technology for industrial symbiosis. *Journal of Industrial Ecology* 14(5): 740–753. doi:10.1111/j.1530-9290.2010.00273.x.

Halstenberg, Friedrich A., Kai Lindow, and Rainer Stark. 2016. Utilization of product lifecycle data from PLM systems in platforms for Industrial Symbiosis. In *Proceedings of the 15th global conference on sustainable manufacturing.*

Hauff, M. von, R. Isenmann, and G. Müller-Christ. 2012. *Industrial ecology management: Nachhaltige Entwicklung durch Unternehmensverbünde:* Springer Gabler.

Herrmann, Christoph, Christopher Schmidt, Denis Kurle, Stefan Blume, and Sebastian Thiede. 2014. Sustainability in manufacturing and factories of the future. *International Journal of Precision Engineering and Manufacturing-Green Technology* 1(4): 283–292.

Jacobsen, Noel B. 2006. Industrial symbiosis in Kalundborg, Denmark: a quantitative assessment of economic and environmental aspects. *Journal of Industrial Ecology* 10(1–2): 239–255.

Kincaid, Judy, and Michael Overcash. 2001. Industrial ecosystem development at the metropolitan level. *Journal of Industrial Ecology* 5(1): 117–126.

Laperrière, L., and G. Reinhart. 2014. *CIRP encyclopedia of production engineering.* New York: Springer.

Laybourn, P. 2015. G7 industrial symbiosis workshop: Faciltiy industrial symbiosis—the circular economy in action.

Laybourn, P., and M. Morrissey. 2009. The pathway to a low carbon sustainable economy: National Industrial symbiosis programme.

Lowe, Ernest. 2007. Eco-industrial park handbook for Asian developing countries. http://teclim.ufba.br/jsf/producaol/indigo%20ecopk%2001_18.PDF.

Ludwig, J. 2012. "Energieeffizienz durch Planung betriebsübergreifender Prozessintegration mit der Pinch-Analyse." Dissertation, Karslruher Institut für Technologie (KIT).

Massard, and Erkmann, eds. 2007. *A regional industrial symbiosis methodology and its implementation in Geneva, Switzerland.*

Massard, and Erkmann, eds. 2009. *A web-GIS tool for industrial symbiosis: Preliminary results and perspectives.*

Mirata, Murat, and Tareq Emtairah. 2005. Industrial symbiosis networks and the contribution to environmental innovation: The case of the Landskrona industrial symbiosis programme. *Journal of Cleaner Production* 13(10): 993–1002.

Nobel, C.E., and D.T. Allen. 2000. Using geographic information systems (GIS) in industrial water reuse modelling. *Process Safety and Environmental Protection* 78(4): 295–303.

Pauli, Gunter. 2010. The blue economy. *Our Planet. GREEN ECONOMY. Making it work. The magazine of the United Nations Environment Programme*, 24–26.

Seliger, Günther. 2008. Sustainable value creation nets. In *Proceedings of the global conference on sustainable product development and life cycle engineering VI, Pusan, Korea*, 2–7.

Shropshire, D.E., D.A. Cobb, P. Worhach, J.J. Jacobson, and S. Berrett. 2000. *Demonstration of decision support tools for sustainable development: An application on alternative fuels in the greater Yellowstone-Teton Region.* Bechtel National Incorporated.

Sokka, Laura. 2011. Local systems, global impacts: Using life cycle assessment to analyse the potential and constraints of industrial symbiosis. Dissertation, University of Helsinki.

Spur, G. 1994. *Handbuch der Fertigungstechnik: Band 6 - Fabrikbetrieb.* München, Wien: Carl Hanser Verlag.

Sterr, Thomas, and Thomas Ott. 2004. The industrial region as a promising unit for eco-industrial development—Reflections, practical experience and establishment of innovative instruments to support industrial ecology. *Journal of Cleaner Production* 12(8): 947–965.

Westkämper, E. 2006. *Einführung in die Organisation der Produktion.* Berlin, Heidelberg, New York: Springer.

Wiendahl, H.-P., J. Reichardt, and P. Nyhuis. 2009. *Handbuch Fabrikplanung. Konzept, Gestaltung und Umsetzung wandlungsfähiger Produktionsstätten.* München: Hanser.

Integration of Sustainability into the Corporate Strategy

Nicole Oertwig, Mila Galeitzke, Hans-Georg Schmieg, Holger Kohl, Roland Jochem, Ronald Orth and Thomas Knothe

Abstract In order to successfully achieve sustainable corporate development, enterprises have to define and implement a pragmatic strategy. In that pursuit, the discussion of motivation and reasoning behind incorporating sustainability strategies serves as a prelude to the thematic examination of challenges and courses of action in corporate strategy development and implementation. Especially in the context of sustainability, additional legislative and stakeholder requirement considerations make managing these tasks effectively, however, much more challenging. The firm's overall objectives thus become multidimensional and have to be broken down to the individual departments and business fields. Consequently, considerable effort has to be devoted to the planning, measurement and evaluation, steering and control as well as optimisation and communication processes of the holistically defined corporate value creation. Furthermore, a solution for enterprise sustainability management and its evaluation is necessary for ultimately balancing economic, ecological and social performance factors, to ensure optimized decision-making.

Keywords Sustainability management · Sustainability strategy · Integrated reporting

1 Organisational Framework for Sustainable Development

With respect to the increasing competitiveness, cost and price pressure as well as the limited availability of natural resources, efficiency—as the maxim of manufacturing—stands as an imperative. Nowadays, a new sense of responsibility

N. Oertwig (✉) · M. Galeitzke · R. Orth · T. Knothe
Fraunhofer IPK, Berlin, Germany
e-mail: nicole.oertwig@ipk.fraunhofer.de

H.-G. Schmieg · H. Kohl · R. Jochem
TU Berlin, Berlin, Germany

© The Author(s) 2017
R. Stark et al. (eds.), *Sustainable Manufacturing*, Sustainable Production,
Life Cycle Engineering and Management, DOI 10.1007/978-3-319-48514-0_12

towards future generations is emerging, as insights on the long-term effects of over-exploitation and environmental pollution are increasing. In the context of the evolution of this responsibility towards internal and external stakeholders, enterprises are confronted with the imminent challenge of adapting strategic orientation and operative value creation accordingly.

The linkage between the economic, ecological and social perspectives of the interaction of enterprises with their environment however, poses unique challenges in terms of potential internal conflicts of objectives. At the same time, it is questionable to what extent the attainment can be related to the three perspectives of sustainability. Thus long-term strategic orientation has to be recognised as a premise for sustainable development, so that potential short-term performance discrepancies are not misinterpreted as deficits, or implied as representing poor decision-making. This is assuming that sustainability is more than an ideological construct for the conscious influence and control of human and entrepreneurial behaviour. Instead, it has to be conditional to certain criteria and traceable or ascertainable. Numerous approaches for operationalising sustainable management are therefore focused on indicators, but remain, however, limited in their extent or integrity in order to avoid complexity.

The three-dimensional differentiated approach requires the simultaneous safeguarding of the economic, ecological and social capacity of the respective system and its environment for both the current and future generations (Dyllick and Hockerts 2002). Building on the definition of the German Bundestag, safeguarding economic performance is herein based on ensuring an adequate competitive situation as a driver of innovation and as a price-building mechanism, without however at the same time limiting the welfare of the individual involved. The preservation, and in some cases, the restoration of the capacity of natural systems, is thus the main objective of the environmental perspective. In that pursuit however, societal order is only sustainable if solidarity and social justice stand as the prerequisites to individual freedom and development in the process of determining the change of conditions and structures (Enquete-Kommission 1998).

Eco-effectivity strategies pursue absolute objectives in terms of reducing environmental pollution, as achieved through the use of renewable energy sources, recirculation of products, by-products and materials into product lifecycles or natural systems, as well as the limitation of environmental pollutants. Eco-effectivity thus refers to the degree of objective attainment, where the target is directly tied to the reduction of environmental or social burdens (Schaltegger 2000).

The fundamental strategy of efficiency is based on the objective of increasing resource productivity through the minimisation of resources deployed in relation to the maximised output with respect to the entire lifecycle. This is commonly achieved through product and process optimisation or innovation as well as procedures and product characteristics profiles that influence the operating condition and lifespan of the product (Enquete-Kommission 1998; OECD 2010). The Eco-efficiency strategy hence refers to resource efficiency in relation to production processes. The substitution of conventional materials—therein enabling the use of less material or the construction of lightweight structures, recyclable materials or those that have lower pollution potential—serves to support the pursuit of eco-efficiency. Socio-efficiency can be expressed in an analogy, wherein value added is expressed in relation to social burden (Schaltegger 2007).

The analysis of a growing world population and simultaneous depletion of natural resources inevitably calls for confrontation with human consumer behaviour (Huber 2011). Sufficiency in an economic context here describes an alignment of consumer behaviour towards a sufficient consumption that accounts for resource depletion with existing technologies. Applied to the organisational level, this entails a limitation of production to a level below the possible growth boundary, so as to avoid overconsumption of natural resources (Huber 2000). The potential for growth of enterprises is not directly limited by the sufficiency strategy. The environmental and social impact is however minimised when implicit consideration of the long-term utilisation of products is taken into account. This represents an attempt at finding an optimal balance between economic value creation and the reduction of environmental pollution and social burden (Bergmann 2010).

Beyond process and product optimisation, the consistency strategy requires a structural change in the utilisation of resources and energy as well as restructured usage of natural drains. This explicitly calls for innovation capability with respect to new technologies, material as well as processes and products (Huber 2011).

This basic model can be extended by four fundamental principles, including responsibility, cooperation, and circular as well as functional orientation. These are possible operational principles held by economic actors, yet are in some cases redundant reiterations of the specifications of strategies and principles on a conceptual level (Dyckhoff and Souren 2008).

From a system theoretical point of view, cause-effect relationships are possible within and between the three dimensions of sustainability. These (inter-) dependencies may be positive or negative, respectively weakening or strengthening effects on the baseline objective of preserving ecological, economic and social capital. The dependencies may be characterised by place, time and reflexivity (Gleich and Gößling-Reisemann 2008). Hence, the effects of actions implemented may appear within the given system currently under consideration or surface in different systems. Simultaneous and delayed effects are often more difficult to detect however, as simultaneous effects may be interpreted as independent, while latent effects may go completely undetected.

2   Incorporating Sustainability Strategies

In order to meet the requirements set forth by the triple bottom line (Dyllick and Hockerts 2002) and the sustainability strategies, enterprises have to adapt their own corporate strategies. In this section, the reasoning behind implementing sustainability as part of the corporate strategy is examined, and the main motivational aspects are highlighted.

While the term strategy stems from a military context (Clausewitz 1935; Giles 1910), the conceptual integration into the context of corporate management in terms of strategic planning and later strategic management, was undertaken over half a century by scholars from varying fields (Will 2012). Originating from conceptions of efficiency as the main driver of productivity (Taylor 1911) and the relation of experience to cost-efficiency (Henderson 1973), competitiveness then took over the corporate strategy discussion, later expounded upon with differentiated business strategies (Porter 1985). The basis for developing a strategy can be dominated by external circumstances such as the market or environment. Moreover, the enterprise typically positions itself through the lens of its internal resource-based perspective —creating value and competitiveness through the deployment of core competencies (Prahalad and Hamel 1990). In that process, a basic definition of strategy as the long-term oriented behaviour of the corporation in pursuit of achieving defined objectives (Welge 2001) needs to be expanded, to account for meeting the corporation's (and its internal and external stakeholders) objectives *together* with safeguarding the same possibility for future generations. In so doing, economic, ecological and social capital have to be expanded, yet sustained for the future (Dyllick and Hockerts 2002).

Based on the historic development of the term and discipline, limitations set forth by sustainability strategies seem contradictory and require closer examination. Initially, the motivational aspects attached to integrating sustainability requirements into the corporate reality are analysed. As for the scientific development of this aspect, a main structuring characteristic lies in the origin of the motivation. Where early contributions were focused on external factors, internal motivation and connecting drivers have gained in significance. Figure 1 gives an overview of the main motivational factors and drivers for corporate sustainability (Bansal and Roth 2000; van Marrewijk and Werre 2003; van Marrewijk 2003; Schaltegger and Burritt 2005; Epstein and Buhovac 2014; Windolph et al. 2014; Lozano 2015; Engert et al. 2016).

Upon consideration of the motivations behind implementing sustainability into the corporate strategy, a new or adapted strategy has to be defined. In a procedural approach to strategy development, the main imperatives and courses of action are discussed in the following section. Here we propose considering the options to (1) adjust the corporate strategy to include objectives regarding economic, ecological and social performance; (2) to define a specific sustainability strategy as part of the corporate strategy and (3) to redefine the corporate strategy based on the premise of creating a holistic sustainability strategy (Figge et al. 2002). After the successful implementation of sustainability aspects in the strategizing phase, proactive management is needed in order to achieve the sustainability objectives.

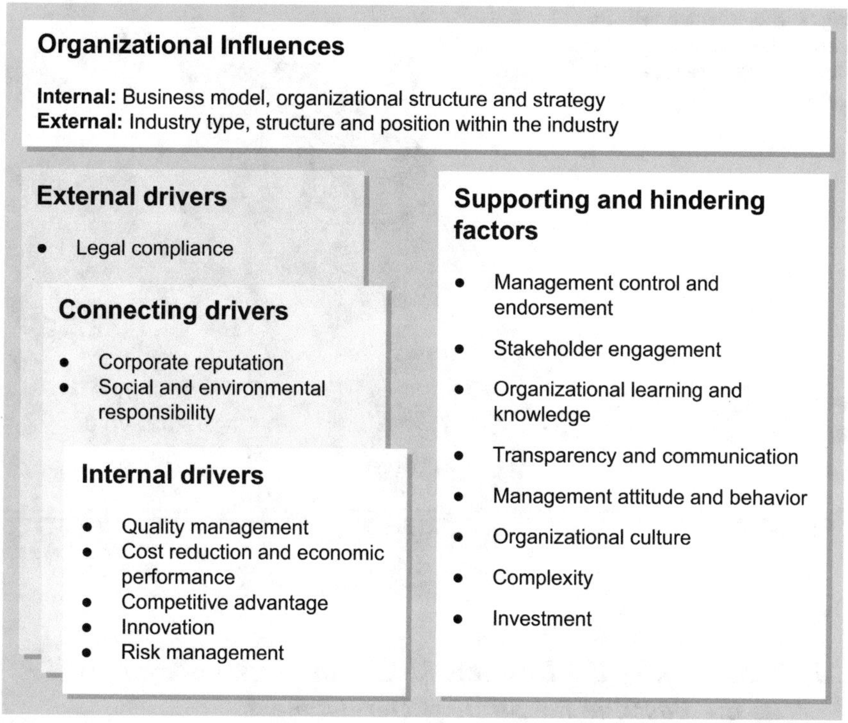

Organizational Influences

Internal: Business model, organizational structure and strategy
External: Industry type, structure and position within the industry

External drivers

- Legal compliance

Connecting drivers

- Corporate reputation
- Social and environmental responsibility

Internal drivers

- Quality management
- Cost reduction and economic performance
- Competitive advantage
- Innovation
- Risk management

Supporting and hindering factors

- Management control and endorsement
- Stakeholder engagement
- Organizational learning and knowledge
- Transparency and communication
- Management attitude and behavior
- Organizational culture
- Complexity
- Investment

Fig. 1 Motivational factors and drivers for corporate sustainability

3   Management of Corporate Sustainability Performance

The management of organisations is described here in a stepwise approach (Fig. 2), addressing the building blocks of the business model, the corporate strategy, the business processes and the resources deployed. In order to improve the performance —in this particular context the sustainability performance—purposeful actions need to be planned, implemented and monitored. Overall, the dynamics of the business operation, decisions taken and the outcome, all need to be recognised in order to establish a comprehensive view of the cause-effect relations within and across the organisation's borders. Communication with relevant stakeholders takes on a key role in that process, as transparency requirements increase. Internal and external communication must become an established activity of organisations that aim to make information available about their performance beyond the standard financial data reporting.

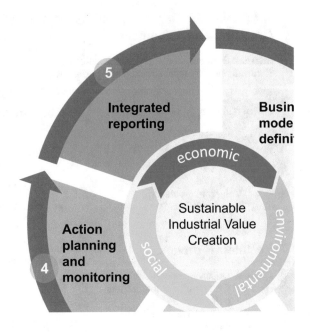

Fig. 2 Stepwise approach for the management of corporate sustainability (Galeitzke et al. 2016)

3.1 Definition of the Business Model and Business Success as the Baseline for Strategy Development

The path of sustainable corporate development needs to be outlined for any business with specific deliberation on its internal and external environment. To achieve sustained success, the organisation must pinpoint its concrete objectives and values. These should be, furthermore, clearly understood, accepted and supported by the employees of the organisation (ISO 2009). It is therefore necessary to explicate the business model and the enterprise's potential innovation as an integral or complementary part of strategy development.

To do adequate justice to the topic of sustainability as a whole, the following perspectives have to be considered within the process of business model definition/innovation:

1. Economic Perspective—While the traditional economic challenges are to increase the company's value and to increase the profitability of products and services, the challenge with regards to economic sustainability lies in making environmental and social management as economical as possible.

2. Environmental Perspective—All actions of an enterprise affect its ecosystem. Thus, companies are encouraged to reduce the absolute level of their negative environmental impact resulting from production processes, products, services, investments etc. to a considerable extent, where the largest possible decrease is desirable. The largest possible decrease is however desirable.

3. Social Perspective—In order to achieve sustainable value creation within the social dimension, the social issues of focus have to provide a real competitive advantage. Such advantages could be obtained by increasing revenues, or reducing risks or operational costs. In this pursuit, the tension between social and economic pressure is relieved as both society and businesses enjoy tangible benefits at the same time.

Combining fragments or modules of a company is a fundamental aspect in several business model definitions (Osterwalder and Pigneur 2011; Johnson et al. 2008; Wirtz 2010; Mitchell and Coles 2003), serving the purpose of creating products and services and thereby creating, providing and maintaining value (Wirtz 2010; Johnson et al. 2008; Osterwalder and Pigneur 2011). In this context, value creation is used for strengthening the customer relationship and competitive advantage (Wirtz 2010). These components of business model innovation can be summarised as illustrated in Fig. 3.

Nowadays innovation is a major key for sustainability due to the fact that the future society demands innovative products, processes and services, without losing out on efficiency (Clausen 2011). Product or incremental process innovations are neither a guarantee for success nor sufficient for coping with the emerging information, knowledge and time-competition (Stern and Jaberg 2010). Against this

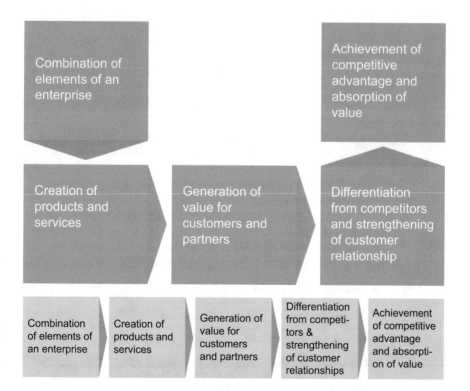

Fig. 3 Constituents of business model innovation definitions (Schallmo 2013)

background, the innovation of business models has arisen as a new discipline, providing organisations with supplementary guidelines for differentiation models in the market place in pursuit of securing long-term competitive advantage. Relating the business model concept to sustainability (Lüdeke-Freund 2010) defines a sustainable business model as "a business model that creates competitive advantage through superior customer value and contributes to a sustainable development of the company and society."

A business model basically defines the way in which a company operates. Sustainable Business model innovation can be an important leverage for change in a company to be considered sustainable and for coping with the emerging challenges in this context. This furthermore entails an expansion of the business model scope beyond green (FORA 2010), product-service-systems (Tukker 2004) or social issues (Yunus et al. 2010; Bocken et al. 2014). Brocken et al. developed a set of sustainable business model archetypes clustered by technological, social and organisational perspective for innovations as shown in Fig. 4 (Bocken et al. 2014).

These archetypes can be interpreted as an approach for business model innovation towards sustainability. They can initially assist in the process of embedding sustainability into existing business models or for the purpose of radical re-engineering of the business models and for delivering a sound starting point from

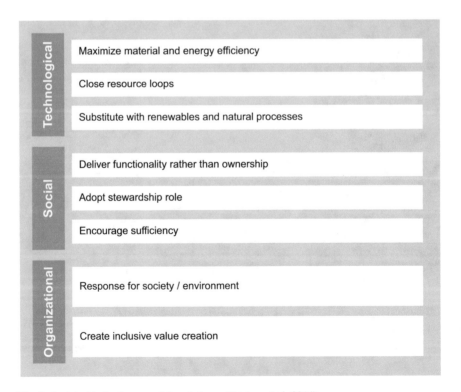

Fig. 4 Sustainable business model archetypes (Bocken et al. 2014)

which to broaden economic, environmental and social aspects in tackling the complementary process of strategy development.

3.2 Strategy Development

Today, enterprises are forced to align their own objectives with the needs of all their stakeholders. Particularly at a time characterised by shorter product life cycles, decreasing prices, new technologies, global markets and increasing sustainability demands, enterprises require an efficient process for their strategy development activities.

The term strategy was first recorded in the late 1950s in the economic doctrine of the Harvard Business School. As instruments of corporate management first evolved from the concept of strategy, the terms strategic planning, and consequently strategic management have been established. In English-speaking countries (Chandler 1962; Ansoff 1965; Schendel and Hofer 1979; Porter 1980), prominent pioneers provided crucial foundations. From this 50-year history of the strategy concept in the context of corporate governance, the following features of a strategy can be derived: the consideration of actions of other actors, proactivity and long-term orientation (Staehle and Conrad 1994).

Strategy in its initial context is generally used to establish conditions that will guarantee long-term economic success and thus the continuity of the company. For this purpose, a strategic success ensues, which ultimately leads to advantages over competitors (Rüegg-Stürm 2005; Grant 2005).

The development of a comprehensive strategy which not only concentrates on competitive benefits and thus on the economic value, presents itself however to be a much more complicated undertaking. With regards to the aspect of sustainability, the environmental and social dimensions have to be taken into account, and, moreover, the cause-impact relations likewise have to be adequately assessed.

Several companies appear to be active in the field of sustainability management. They may publish, for example, extensive sustainability reports. Yet their efforts nevertheless often remain unclear from a strategic perspective. Rather, the impression that sustainability issues are being tracked often tends to be the case, more than they are actually proceeding on the basis of a clear strategy (Baumgartner and Ebner 2010).

The development of a comprehensive enterprise strategy which meets all given requirements from internal and external stakeholders and specifically contains sustainability perspectives, is a process which requires a structured approach in the interest of keeping the complexity and uncertainty at a minimum level. The process of strategy development can be divided into four major phases as presented in Fig. 5 (Will 2012).

In the first step, information is preliminarily collected which describes the current situation of the company for establishing a general consensus on the initial

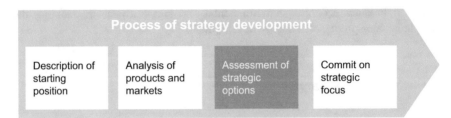

Fig. 5 Strategy development

situation (e.g. information about business environment, general corporate objectives or the corporate profile incl. development of earnings).

In the second step, the products and markets are categorised so as to quantify their respective contribution toward the overall business result. For visualisation, the findings can be represented e.g. in a product-market-chart. Based on this analysis, the current market situation of the company is evaluated. The aim of this step is to obtain a first rough estimation of the yield model to derive interesting advancements from the existing business model in the next step.

The major decisions regarding the incorporation of sustainability into the strategic decision-making process are derived in the step of assessing the strategic options for corporate sustainability. The starting point for the determining of suitable strategic options is captured in step 1, featuring the general corporate objectives and the current trends in the business environment. In addition, the current situation of the company examined in step 2 leads to the necessity of a fundamental decision on how exactly the company would like to deal with the challenge of sustainability without losing any growth potential. Baumgartner and Ebner (2010) recommend a set of profiles for sustainability strategy (Table 1) as a first means of orientation in the strategic decision-making process.

Each of these positions the company wants to occupy has to be evaluated by taking into account risks, chances and possible development scenarios regarding market penetration, product differentiation, market expansion or diversification. For the analysis of the relationship between sustainability and competitive strategy, (Baumgartner and Ebner 2010) propose two criteria: costs caused by the strategy, and the recipient of the resulting benefits.

Finally, a selection of a strategic option based on the assessment from the previous step has to take place in order to arrive at the detailed strategic objective as a conclusion.

Since an enterprise consists of several different units and elements which are interconnected on several levels (active vs. passive or strong vs. weak relationship), it is necessary to consider all influences and possible side-effects within the process of strategy implementation. In this context, many companies use enterprise processes as a common backbone for the different management disciplines with the objective of fast and consistent realisation of strategic issues at all levels of the enterprise (Jochem and Balzert 2010).

Table 1 Strategy profiles for sustainability based on Baumgartner and Ebner (2010)

Strategy profile	Explanation
Introverted	• Low standard of sustainability • Concentrates mainly on conformity and compliance with sustainability rules and guidelines
Conventional extroverted	• Aims to communicate sustainability commitment to society for increasing competiveness • Responsibility often located in public relation department • Focused on external presentation of sustainability
Transformative extroverted	• General orientation conventionally extroverted • Company is a driver for corporate sustainability in society • Most important are facts, which prompt sensitive reaction from society without proving fulfilment
Conservative	• Oriented towards internal measures • Focusing cost efficiency and well defined processes • Commitment to investment in appropriate technology, sophisticated health and safety, ecological sustainability • Process-based analysis and assessment of corporate sustainability • Society-related issues less important
Systemic visionary	• Highly developed sustainability commitment • Combines outside-in and inside-out perspective, based on internalisation and continuous improvement of sustainability issues • Aims in all sustainability aspects at good results • Stakeholders and market are equally addressed by sustainability commitment
Conventional visionary	• Oriented towards market impact • High level of maturity • Minimal lower maturity in processes, purchasing, no controversial activities or corporate citizenship due to lower impact to market situation as sustainability leader

The use of process management approaches for transferring complex strategies down to the operational business will be examined in the following section.

3.3 Process Definition and Modelling

Process definition and modelling is of great importance in the pursuit of achievement of the company's strategic and operational objectives. The aim is to improve the efficiency on the one hand, and the effectiveness of the company on the other hand, so that its total value can be increased. Processes and process management are connected to two essential signifiers for ensuring effectiveness and efficiency in the company. First, the corporate strategy determines the processes which are required and which strategic objectives are to be implemented alongside them. It forms the basis for process identification and target orientation. This involves changes in corporate strategy, entailing changes in the processes itself. Secondly, the customer or stakeholder orientation determines what expectations and requirements have to be met through the processes. Therefore, the process definition extends from the requirements of the customer to the delivery of the process results to the client. It is important that the terms of the processes of corporate strategy and customer reference in the context of process management are coordinated (Jochem and Balzert 2010). Figure 6 illustrates the connection of corporate strategy and its operationalisation via an integrated management.

The comprehensive development and implementation of a corporate sustainability strategy which meets the requirements of the economic, environmental and social perspective, require a sound information basis from which to proceed. The various management disciplines involved have to be addressed in such a way that the attendant complexity is reduced to a minimum. A promising approach for visualizing and therein explaining the interrelation of varied enterprise objects lies in enterprise modelling.

In Vernadat's view (1996), an enterprise model is the basis for the understanding of a company, whereby the relevant structural and dynamic components and their interactions are described.

Enterprise modelling describes relevant processes and structures of a company or organisation and their mutual relationships. The applications are designed extend to the illustration of the enterprise architecture, the root cause analysis of

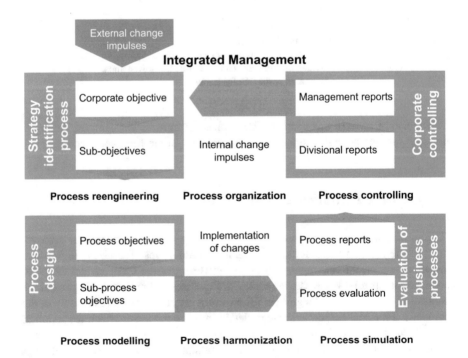

Fig. 6 Connection of corporate strategy and process management

operational problems, strategy development, process optimisation or the management of business collaborations, among other topics (Sandkuhl et al. 2013).

Thus, the process management commences with the alignment of the processes and the sustainability strategy, which means defining the value-adding processes and objectives to be achieved. In the following phase of process design, the defined processes will be designed in detail, modelled and optionally documented. In the course of the implementation of the processes in the organisation of the company, the evaluation of the processes is carried out in terms of target-achievement, and where applicable, harmonisation or standardisation can be required. Finally, the actual controlling of processes follows, related to the entire corporate controlling process, resulting in impacts on the strategic development.

Both the challenges and the opportunities which integrated mapping of process management and sustainability offers, lie mainly in the mastery of increasingly complex planning processes. Based on enterprise models that unite the perspectives of different strategic planning disciplines and also support them with integrated model-based planning and evaluation instruments, the objective of corporate sustainability is pursued holistically (Dyllick and Hockerts 2002).

An important and critical success factor remains however unconsidered within enterprise models. The implementation of a sustainable development strategy requires not only an excellent knowledge of the internal processes and structures, but also, for example, of relationships with customers and partners, i.e. intangible assets. The role of such assets in terms of sustainability is briefly introduced in the next section, along with an approach for the integration of these values into the development of corporate sustainability.

3.4 Resource Definition and Impact Analysis

In order to provide products or services, an organisation will combine different types of resources like human skills and knowledge, natural materials and social structures, by using machinery, infrastructures and financial assets. A sustainable organisation will maintain and, wherever possible, enhance these capital assets, rather than exhaust them ("capital stewardship") (Knight 2006; ARE and DEZA 2004). In turn, the design of the business processes constitutes the interrelation of the business operation, its resources and performance as well as the impact on the economic, social and environmental dimensions (Fig. 7). If, for instance, economic sustainability is interpreted as an expansion of the private welfare maximisation, enterprises have to ensure the long-term functionality and effective performance of their operation. Consequently, the design of the business processes needs to be directed towards the effective, efficient and beneficial use as well as towards the development of the capital assets involved. In this context, the capital-based approach refers to the relevance of different types of resources and makes a basic distinction between tangible and intangible resources. These are then employed in business processes to improve the organisational performance.

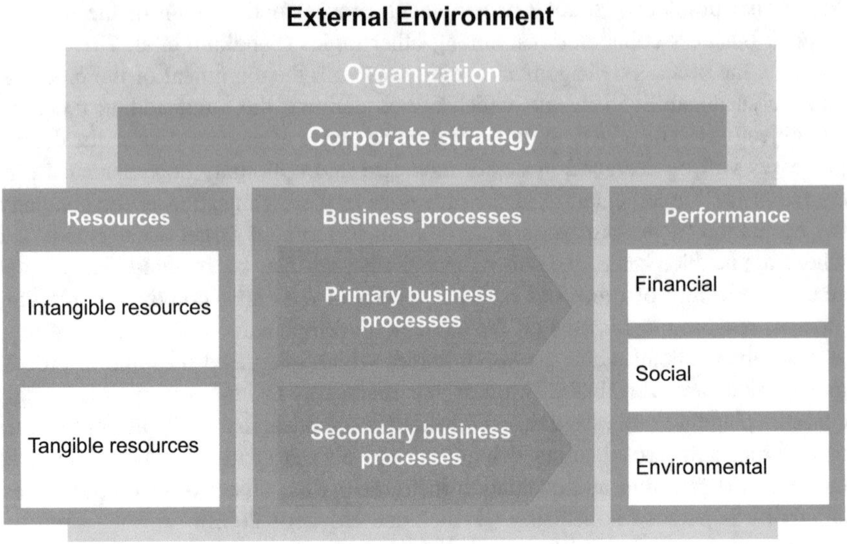

External Environment

Fig. 7 Reference model for corporate sustainability

Tangible resources, meaning those resources that are material or substantial, are composed of financial, manufactured and natural capital (IIRC 2013).

Financial capital is the sum of available financial resources that are utilised to fund the organisation's operation. Thus, the product and service provisions are financially sustained through capital obtained via revenues, investments, debt, equity or grants.

Manufactured capital meanwhile comprises all physical objects that are employed by the organisation in order to produce and deliver its products and services. This physical part of the production system includes infrastructure and buildings, operating equipment as well as measuring, storage and transport utilities (Westkämper and Decker 2006). These objects can be obtained from third parties or in-house production.

On the basis of the classical understanding of "land" as a major factor of production, natural capital comprises all natural resources, processes and systems available (Harris and Roach 2013; IIRC 2013).

The classification of intellectual capital as an intangible resource follows the principle of the harmonisation of intellectual capital factors into standard repositories. Human, structural and relational capital are herein subdivided into standard success factors (Mertins and Will 2008) which map the most common types of intellectual capital. In order to comply with the system attached to modelling processes, the repository of intellectual capital factors needs to be adapted on a case-by-case basis. At the same time, considerations for directing this approach towards sustainable corporate development are taken in the following adaptation delineation.

The competence model forms the basis for the human capital factors. It was developed through empirical studies and quantifies specifics of enterprises analysed. Here a more generic approach is taken, which in turn is detailed through the consideration of role- and activity-based competencies. Human capital is thus defined as the sum of professional, social, personal and methodological competence. The peculiarity of these competences is dependent on the specific role occupied or on the activity itself, and in a wider sense, likewise on the strategic consideration of paradigms such as sustainable development.

The structural capital requires a distinct consideration of those capital factors that are activity-based (cooperation and knowledge transfer, product and process innovation), and the objectified factors (management instruments, explicit knowledge and corporate culture). While all factors are indeed structural factors of intangible resources, the implications on the activities of the model as condition transformation of objects such as "knowledge," need to be observed and incorporated into the process model creation.

In relational capital, a new configuration considers relations on micro-, meso- and macro-level in order to integrate social aspects in a distinguished manner. At the micro level, the external relationships of the enterprise with individual actors are considered, while cooperation partners, supplier-, customer- and investor-relationships constitute the meso-level as individual "dyadic" relationships (Provan et al. 2007). Relationships to public bodies (legislative, funding) and society moreover are considered within the macro-level of relational capital. This allows for a focused definition of all relevant stakeholders and the enterprise's relationships to those stakeholders.

At this point, an assessment of the cause-effect relationships can be implemented following a cross-factor impact assessment of all resource factors (Alwert et al. 2005). Identifying closed-loop interrelations is an attempt to address the system's theoretical discussion of the introduction, where weakening or strengthening dependencies are identified and expressed in relation to a specific analysis object (Galeitzke et al. 2015).

The definition of resources (tangible and intangible) builds the basis for analysing the interrelations within the different resource categories and helps to identify fields of action for improving on the sustainability performance of their deployment. The following section introduces an approach for action planning and monitoring by using extended enterprise models.

3.5 Action Planning and Monitoring Through Allocation in Process Models

The most brilliant sustainability strategies can turn into disasters if they are not entirely or only insufficiently implemented. A key factor for a successful implementation of the sustainability strategy lies in the planning of operational actions and the availability of evaluations for monitoring and tracking qualitative and quantitative aspects. The measurement, control and communication of information on sustainability require the interaction between various actors, evaluation methods and operational data (Maas et al. 2016).

Figure 8 presents a framework concept for the description, analysis and monitoring of sustainability, specifically their interrelation with enterprise models.

Applying this framework, one can ensure that a systematic embedding of the individual sustainability strategies, objectives, their monitoring and its implementation takes place in the planning phase.

The enterprise model characterises the core area of the framework presented. It represents an enterprise within all its aspects of strategic objectives, products, organisation, processes, tangible and intangible resources and their interrelation to each other. Once the variables that contribute to the characterisation of sustainability are modelled, a detailed action plan for the achievement of the strategic objectives is required. In order to coordinate this multi-dimensional sustainability system, mechanisms for prioritising them, clustering mechanisms for mapping them to the different dimensions of sustainability, as well as mechanisms for describing the relation aspects between them, are all necessary. To make best use of the scarce resources of an enterprise, an initial selection is necessary. To that end, a two-dimensional prioritisation-matrix can be used. The matrix differentiates between the dimensions "need for action (urgency)" and "feasibility"—each of them assuming the characteristic values low, medium and high. The matrix (Fig. 9) can help identify which measures are urgent and how easy or difficult they are to implement (Kohl et al. 2014).

It reveals the urgency level of the actions, along with their feasibility. The optimisation of the energy use might, for example, be highly urgent, but need not be easily feasible due to contractual ties. Furthermore, the enhancement of the material efficiency could be highly urgent, but not very feasible, due to the complex processes along the value chain that can only be altered with the application of enormous effort.

As soon as the prioritisation is complete, a suitable set of indicators has to be derived. Due to that fact, numerous methods, guidelines and norms have been developed (Kohl et al. 2013; Neugebauer et al. 2015; ISO 2013; VDI 2016), which offer evaluation mechanisms, and finally, indicators for expressing the degree of target achievement. A further consideration is then omitted at this point. Once the suitable indicators are aligned with the planned actions and thus with the strategic objectives, the monitoring via the usage of operational data has to be realised. Business intelligence and reporting tools that are only capable of visualising performance indicators are no longer sufficient for capturing the complex requirements of a comprehensive sustainability approach (Schneider and Meins 2012). Moreover, a solution for network sustainability management and its evaluation is required for balancing economic, ecological and social dimensions (Wilding et al. 2012). In the context of sustainable development, economic, environmental and social aspects have to be presented in a context-sensitive manner. To provide task or role-oriented information, the framework supports a so-called "view concept." The views contain

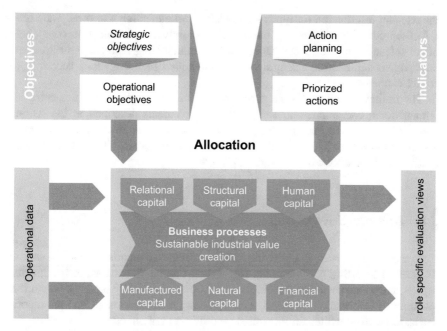

Fig. 8 Model-based framework the management of corporate sustainability performance

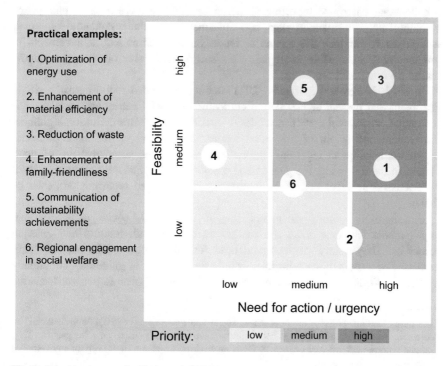

Fig. 9 Prioritisation-matrix (Orth et al. 2011)

the relevant information for typical application and modelling purposes. They offer a focused cut without changing the models themselves. An evaluation component offers role-specific model evaluation views, summarizing relevant indicators and enterprise information in a central system, and allows their evaluation according to model elements.

The framework also allows a derivation of integrated reporting which complies with national and international standards. All elements described in the section above and integrated into the integrated model-based framework, are represented also in reporting guidelines for the communication of sustainability. The following section briefly introduces the major approaches.

3.6 Integrated Reporting

Companies are exposed to a growing number of required reports for internal as well as external reporting purposes (e.g. Intellectual-Capital-Statements, environmental reports, corporate social responsibility reports or sustainability reports). Given this situation of information overload, a comprehensive integration of various reports seems to be worthwhile. An integrated reporting format would not only reduce the internal preparation efforts, but also contribute to the standards, as for example formulated in the EU directive "Accounts Modernization Directive" on non-financial enterprise reporting (Clausen et al. 2006). While large enterprises communicate non-financial data and information to their stakeholders, small enterprises so far lack the means to report on their effort and achievements in implementing sustainable strategies. This section highlights our research contribution on integrated reporting.

In 2011, Eccles and Saltzman (2011) defined integrated reporting as "a single document that present and explain a company's financial and nonfinancial—environmental, social, and governance (ESG)—performance." This definition highlights the content and origin of integrated reports. In addition to traditional financial information, contents regarding the sustainability of the company[1] are of note. Hence, in the following, the phenomena surrounding "sustainability reporting" will be discussed in detail before the connection to integrated reporting will then be drawn.

Sustainability reports document the environmental, social and economic engagements that enterprises are making in dealing with internal and external resources. They satisfy the increased need for information on the part of stakeholders. For sustainability reporting, criteria and an array of guidelines are already available. Worldwide attention has been paid to the Global Reporting Initiative

---

[1]The terms "sustainability", "environmental, social and governance" (ESG), "non-financial" or "corporate social responsibility" (CSR) reporting are frequently used interchangeably. They describe reports with different degrees of focus on environmental, social or corporate governance issues (Ioannou and Serafeim 2011).

(GRI). Since 2013, the meanwhile fourth version of the so-called "G4 Guidelines"—is available (Global Reporting Initiative 2013). Since the so-called "CSR directive" of the European Union was released, all reports published after the 6th of December 2016 have to be prepared "in accordance" with the G4-Guidelines (Guideline 2014/95/*EU*). When developing the guidelines, the GRI had several objectives in mind. One was to offer a bridge-builder for sustainability reporting on the path toward integrated reporting. The G4-Guidelines are therefore also applicable and implementable in integrated reporting (Soyka 2013).

The International Integrated Reporting Council (IIRC)—established in August 2010—consists of representatives from corporate, investment, accounting, securities, regulatory, academic and standard-setting sectors as well as from civil society (IIRC 2011). In September 2011, the IIRC released its first discussion paper, offering an initial proposal for the development of an "International Integrated Reporting Framework." More than 200 responses were received from a wide range of stakeholder groups. The (IIRC 2012) published the results in 2012. The current IIRC proposal considers arguments for integrated reporting, and describes guiding principles and content while offering preliminary suggestions for the development of an international "integrated reporting framework" (IIRC 2013).

Central to Integrated Reporting is the organisation's business model, i.e. "the process by which an organisation seeks to create and sustain value" in the short-, medium- and long-term perspective. This model is embedded into a system of inputs, business activities (the core of the business model) and outputs, as well as outcomes. In this context, value creation is not done by or within the organisation alone. It is influenced by external factors, e.g. the economic conditions and societal issues which represent risks and opportunities in the external environment. Furthermore, relationships to employees, partners, networks, suppliers and customers have an impact on the organisation's value creation process. All organisations depend on different resources and relationships for their success. In that process, the IIRC framework uses the concept of "multiple capitals" for explaining how an organisation creates and sustains value. According to the framework, an integrated report should display an organisation's stewardship not only with regards to financial capital, but also with other forms of "capital" (e.g. manufactured, human, intellectual, natural and social), along with their interdependencies.

According to the IIRC, integrated reporting explains linkages between an organisation's strategy, governance and financial performance and the social, environmental and economic context within which it operates. Based on this, the IIRC formulates suggestions for integrated reporting—consisting of seven guiding principles and nine key content elements. The guiding principles underpin the preparation of an integrated report, based on the interconnected key content elements.

The Guiding Principles are:	The Content elements are:
A. Strategic focus and future orientation	A. Organisational overview and external environment
B. Connectivity of information	B. Governance
C. Stakeholder relationships	C. Business model
D. Materiality	D. Risks and opportunities
E. Conciseness	E. Strategy and resource allocation
F. Reliability and completeness	F. Performance
G. Consistency and comparability	G. Outlook
	H. Basis of preparation and presentation
	I. General reporting guidance

The approach of the IIRC gives comprehensive understanding of tangible and intangible resources and suggests interdependencies between corporate action and results. Since the IIRC approach aims for a harmonisation of reporting, a special focus is set on the enterprise's external communication.

Originally, the approach was developed for large companies that are publicly traded. However, an approach for small- and medium-sized enterprises (SME) must be "downsized" or "downsizable" for the special purposes of SME (Bornemann

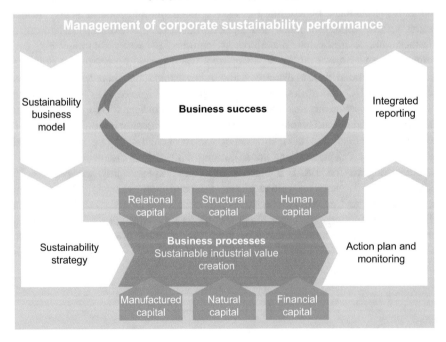

Fig. 10 Framework for management of corporate sustainability performance

et al. 2011). Because the IIRC approach principle is based on this, flexibility for an adaption is thus built-in.

In-line with the guiding principles and content of the IIRC, the authors have developed a reduced approach with a special focus on SME. This approach uses the five following principles and six content suggestions:

The Guiding Principles are:	The Content elements are:
A. Materiality	A. Organisational overview
B. Integrity	B. External environment
C. Connectivity	C. Business model
D. Consistency and comparability	D. Risks and opportunities
E. Communicative quality	E. Performance
	G. Actions and Outlook

To enhance the range in the distribution of the report, the approach also suggests using digital media. In addition, the formulated principles likewise profit from the use of digital media. When regarding, for instance, the consistency and comparability principle, the timelines of the KPIs prove to be much more doable in the digital approach than in the case of a classical print-version of a report.

4 Conclusion

The proposed integrated model-based framework for the management of corporate sustainability performance and the presented stepwise approach for implementing the discussed elements can be summarised as illustrated in Fig. 10. It can assist researchers as well as practitioners in gaining a clearer focus on the development and implementation of sustainability business models, sustainability strategies, performance management and reporting, regardless of whether transparency or decision support is taken as an a priori perspective. It also enables managers to improve their understanding of how the different management disciplines interact on sustainability topics and how to tackle increasing complexity in a context-sensitive and role-based concept.

Further steps in the area of sustainability performance management are nevertheless needed to extend the scope towards complete supply chains in order to manage, evaluate and control the performance of complex value-creation networks. Here, detailed concepts for an intuitive handling of data occurrence means that services for its selection, combination and aggregation, all have to be examined. In addition, several evaluation methods like the LCA already exist on the market, but connection mechanisms have to be developed to allow for reliable steering, controlling and monitoring. On top of the data-driven development needs, the knowledge transfer to the industrial community also has to be strengthened in order to improve and support the corporate sustainability orientation process as a whole.

References

Alwert, Kay, Peter Heisig, and Kai Mertins. 2005. *Wissensbilanzen — Intellektuelles Kapital erfolgreich nutzen und entwickeln*. Berlin: Springer.

Ansoff, H.I. 1965. *Corporate strategy: An analytic approach to business policy for growth and expansion*. New York: McGraw-Hill.

ARE, and DEZA. 2004. Nachhaltige Entwicklung in der Schweiz: Methodische Grundlagen. http://www.are.admin.ch/themen/nachhaltig/00260/index.html?lang=de. Accessed 20 June 2016.

Bansal, Pratima, and Kendall Roth. 2000. Why companies go green: a model of ecological responsiveness. *Academy of Management Journal* 43(4): 717–736. doi:10.2307/1556363.

Baumgartner, Rupert J., and Daniela Ebner. 2010. Corporate sustainability strategies: Sustainability profiles and maturity levels. *Sustainable Development* 18(2): 76–89. doi:10. 1002/sd.447.

Bergmann, Lars. 2010. *Nachhaltigkeit in Ganzheitlichen Produktionssystemen*. Schriftenreihe des Instituts für Werkzeugmaschinen und Fertigungstechnik der TU Braunschweig. Essen: Vulkan-Verl. Techn. Univ., Diss.—Braunschweig, 2009.

Bocken, N.M.P., S.W. Short, P. Rana, and S. Evans. 2014. A literature and practice review to develop sustainable business model archetypes. *Journal of Cleaner Production* 65: 42–56. doi:10.1016/j.jclepro.2013.11.039.

Bornemann, Manfred, Kay Alwert, and Ronald Orth. 2011. Comments on the discussion paper Towards Integrated Reporting—Communicating Value in the 21st Century.

Chandler, Alfred D. 1962. *Strategy and structure: Chapters in the history of the industrial enterprise*. Cambridge: MIT Press.

Clausen, Jens. 2011. Mit Innovationen auf den Weg zur Nachhaltigkeit. http://www. globalcompact.de/sites/default/files/jahr/publikation/1a_clausen_jens_-_mit_innovationen_ zur_nachhaltigkeit_-_borderstep_institut.pdf. Accessed 21 June 2016.

Clausen, Jens, Thomas Loew, and Walter Kahlenborn. 2006. Lagebericht zur Lageberichterstattung. http://www.umweltbundesamt.de/sites/default/files/medien/publikation/ long/3236.pdf. Accessed 20 June 2016.

Clausewitz, Carl v. 1935. *Vom Kriege*. Leipzig: Quelle & Meyer.

Dyckhoff, Harald, and Rainer Souren. 2008. *Nachhaltige Unternehmensführung: Grundzüge industriellen Umweltmanagements; mit 13 Tabellen*. Berlin: Springer.

Dyllick, Thomas, and Kai Hockerts. 2002. Beyond the business case for corporate sustainability. *Business Strategy and the Environment* 11(2): 130–141. doi:10.1002/bse.323.

Eccles, Robert G., and Daniela Saltzman. 2011. Achieving sustainability through integrated reporting. *Stanford Social Innovation Review* (Summer 59).

Engert, Sabrina, Romana Rauter, and Rupert J. Baumgartner. 2016. Exploring the integration of corporate sustainability into strategic management: A literature review. *Journal of Cleaner Production* 112: 2833–2850. doi:10.1016/j.jclepro.2015.08.031.

Enquete-Kommission. 1998. Innovationen zur Nachhaltigkeit.

Epstein, Marc J., and Adriana R. Buhovac. 2014. *Making sustainability work: Best practices in managing and measuring corporate social, environmental, and economic impacts*. San Francisco: Berrett-Koehler Publishers.

Figge, Frank, Tobias Hahn, Stefan Schaltegger, and Marcus Wagner. 2002. The sustainability balanced scorecard—linking sustainability management to business strategy. *Business Strategy and the Environment* 11(5): 269–284. doi:10.1002/bse.339.

FORA. 2010. Green business models in the Nordic Region: A key to promote sustainable growth. http://www.danishwaterforum.dk/activities/Water_and_green_growth/greenpaper_fora_ 211010_green_business%20models.pdf. Accessed 20 June 2016.

Galeitzke, Mila, Nicole Oertwig, Ronald Orth, and Holger Kohl. 2016. Process-oriented design methodology for the (inter-) organizational intellectual capital management. *Procedia CIRP* 40: 674–679. doi:10.1016/j.procir.2016.01.153.

Galeitzke, Mila, Erik Steinhöfel, Ronald Orth, and Holger Kohl. 2015. Strategic intellectual capital management as a driver of organisational innovation. *International Journal of Knowledge and Learning* 10(2): 164–181. doi:10.1504/IJKL.2015.071622.

Giles, Lionel. 1910. *Sun Tzu on the art of war: The oldest military treatise in the world.* London: Luzac & Co.

Gleich, Arnim v., and Stefan Gößling-Reisemann. 2008. *Industrial Ecology: Erfolgreiche Wege zu nachhaltigen industriellen Systemen.* Wiesbaden: Vieweg + Teubner Verlag.

Global Reporting Initiative. 2013. Sustainability Reporting Guidelines: Version 4.0. https://www. globalreporting.org/standards/g4/Pages/default.aspx. Accessed 20 June 2016.

Grant, Robert M. 2005. *Contemporary strategy analysis,* 5th ed. Malden, Mass. [u.a.]: Blackwell.

Harris, Jonathan M., and Brian Roach. 2013. *Environmental and natural resource economics: A contemporary approach,* 3rd ed. Armonk: M.E. Sharpe.

Henderson, Bruce C. 1973. *The Experience curve-reviewed (Part IV): the growth share martrix of the product Portfolio.* Boston: Boston Consulting Group.

Huber, Joseph. 2000. Industrielle Ökologie. Konsistenz, Effizienz und Suffizienz in zyklusanalytischer Betrachtung. In *Global change - globaler Wandel: Ursachenkomplexe und Lösungsansätze; causal structures and indicative solutions,* ed. Rolf Kreibich and Udo E. Simonis. Wissenschaft in der Verantwortung. Berlin: Arno Spitz.

Huber, Joseph. 2011. *Allgemeine Umweltsoziologie.* 2., vollst. überarb. Aufl. Wiesbaden: VS Verlag für Sozialwissenschaften.

IIRC. 2011. Towards Integrated Reporting: Communicating Value in the 21st Century. http:// integratedreporting.org/wp-content/uploads/2011/09/IR-Discussion-Paper-2011_spreads.pdf. Accessed 20 June 2016.

IIRC. 2012. Towards Integrated Reporting—Communicating Value in the 21st Century: Summary of Responses to the Summar of Responses to the September 2011 Discussion Paper and Next Steps. http://integratedreporting.org/wp-content/uploads/2012/06/Discussion-Paper-Summary1. pdf. Accessed 20 June 2016.

IIRC. 2013. The International <IR> Framework. http://integratedreporting.org/wp-content/ uploads/2015/03/13-12-08-THE-INTERNATIONAL-IR-FRAMEWORK-2-1.pdf. Accessed 20 June 2016.

Ioannou, Ioannis, and George Serafeim. 2011. The consequences of the consequences of mandatory corporate sustainability reporting. *Harvard Business School Working Paper 11– 100.*

ISO. 2009. *Leiten und Lenken für den nachhaltigen Erfolg einer Organisation - Ein Qualitätsmanagementansatz,* no. ISO 9004:2009. Berlin: Beuth.

ISO. 2013. *Environmental management—environmental performance evaluation - Guidelines,* no. DIN EN ISO 14031. Berlin: Beuth.

Jochem, Roland, and Silke Balzert (eds.). 2010. *Prozessmanagement: Strategien, Methoden, Umsetzung,* 1st ed. Düsseldorf: Symposion Publishing.

Johnson, Mark W., Clayton M. Christensen, and Henning Kagermann. 2008. Reinventing your business model. *Harvard Business Review* 86(12): 50–59. doi:10.1225/R0812C.

Knight, Dave. 2006. The SIGMA management model. In *Management models for corporate social responsibility,* ed. Jan Jonker, and Marco de Witte, 11–18. Berlin: Springer.

Kohl, Holger, Ronald Orth, and Oliver Riebartsch. 2013. Sustainability analysis for indicator-based benchmarking solutions. In *11th Global Conference on Sustainable Manufacturing, GCSM 2013// Innovative solutions: Abstracts,* ed. Günther Seliger, 567– 573. Berlin: Universitätsverlag der TU Berlin.

Kohl, Holger, Ronald Orth, and Erik Steinhöfel. 2014. Process-oriented knowledge management in SMEs. In *Proceedings of the 15th European Conference on Knowledge Management (ECKM),* ed. C. Vivas and P. Sequeira, 563–570. Academic Conferences and Publishing International Limited.

Lozano, Rodrigo. 2015. A holistic perspective on corporate sustainability drivers. *Corporate Social Responsibility and Environmental Management* 22(1): 32–44. doi:10.1002/csr.1325.

Lüdeke-Freund, Florian. 2010. Towards a conceptual framework of 'business models for sustainability'. In *ERSCP-EMSU Conference, 2010, in Delft.*

Maas, Karen, Stefan Schaltegger, and Nathalie Crutzen. 2016. Integrating corporate sustainability assessment, management accounting, control, and reporting. *Journal of Cleaner Production.* doi:10.1016/j.jclepro.2016.05.008.

Mertins, Kai, and Markus Will. 2008. Strategic relevance of intellectual capital in European SMEs and sectoral differences. InCaS: Intellectual Capital Statement—Made in Europe. In *Proceedings of the 8th European Conference on Knowledge Management, Barcelona.*

Mitchell, Donald, and Carol Coles. 2003. *The ultimate competitive advantage: Secrets of continually developing a more profitable business model*, 1st ed. San Francisco: Berrett-Koehler.

Neugebauer, Sabrina, Julia Martinez-Blanco, René Scheumann, and Matthias Finkbeiner. 2015. Enhancing the practical implementation of life cycle sustainability assessment—proposal of a Tiered approach. *Journal of Cleaner Production* 102: 165–176. doi:10.1016/j.jclepro.2015.04. 053.

OECD. 2010. *Eco-innovation in industry: Enabling green growth*, 1st ed. Paris: OECD.

Orth, Ronald, Stefan Voigt, and Ina Kohl. 2011. *Praxisleitfaden Wissensmanagement: Prozessorientiertes Wissensmanagement nach dem ProWis-Ansatz einführen.* Stuttgart: Fraunhofer.

Osterwalder, Alexander, and Yves Pigneur. 2011. *Business model generation: Ein Handbuch für Visionäre, Spielveränderer und Herausforderer.* Frankfurt am Main: Campus.

Porter, M.E. 1985. *Competitive advantage: Creating and sustaining superior performance.* New York: Free Press.

Porter, Michael E. 1980. *Competitive strategy: Techniques for analyzing industries and competitors.* New York: Free Press.

Prahalad, Coimbatore K., and G. Hamel. 1990. The core competence of corporation. *Harvard Business Review* 69(4): 81–92.

Provan, Keith G., Amy Fish, and Joerg Sydow. 2007. Interorganizational networks at the network level: A review of the empirical literature on whole networks. *Journal of Management* 33(3): 479–516. doi:10.1177/0149206307302554.

Rüegg-Stürm, Johannes. 2005. *Das neue St. Galler Management-Modell: Grundkategorien einer integrierten Managementlehre; der HSG-Ansatz.* 2., durchges. Aufl., [Nachdruck]. Bern: Haupt.

Sandkuhl, Kurt, Matthias Wißotzki, and Janis Stirna. 2013. *Unternehmensmodellierung: Grundlagen, Methode und Praktiken.* Berlin: Springer.

Schallmo, Daniel. 2013. *Geschäftsmodell-Innovation: Grundlagen, bestehende Ansätze, methodisches Vorgehen und B2B-Geschäftsmodelle.* Wiesbaden: Springer Fachmedien.

Schaltegger, Stefan (ed.). 2000a. *Wirtschaftswissenschaften. Studium der Umweltwissenschaften.* Berlin: Springer.

Schaltegger, Stefan, ed. 2000. 2007. *Nachhaltigkeitsmanagement in Unternehmen: Von der Idee zur Praxis: Managementansätze zur Umsetzung von Corporate Social Responsibility und Corporate Sustainability.* Berlin: Bundesministerium für Umwelt Naturschutz und Reaktorsicherheit (BMU) Referat Öffentlichkeitsarbeit.

Schaltegger, Stefan, and Roger Burritt. 2005. Corporate sustainability. In *The international yearbook of environmental and resource economics 2005/2006: A survey of current issues*, ed. Thomas H. Tietenberg and Henk Folmer, 185-. New horizons in environmental economics series. Cheltenham: Edward Elgar.

Schendel, Dan, and Charles W. Hofer. 1979. *Strategic management: A new view of business policy and planning.* Boston: Little, Brown.

Schneider, Anselm, and Erika Meins. 2012. Two dimensions of corporate sustainability assessment: Towards a comprehensive framework. *Business Strategy and the Environment* 21(4): 211–222. doi:10.1002/bse.726.

Soyka, Peter A. 2013. The International Integrated Reporting Council (IIRC) integrated reporting framework: Toward better sustainability reporting and (Way) beyond. *Environmental Quality Management* 23(2): 1–14. doi:10.1002/tqem.21357.

Staehle, Wolfgang H., and Peter Conrad. 1994. *Management: Eine verhaltenswissenschaftliche Perspektive.* 7. Aufl./ überarb. von Peter Conrad. Vahlens Handbücher der Wirtschafts- und Sozialwissenschaften. München: Vahlen.

Stern, Thomas, and Helmut Jaberg. 2010. *Erfolgreiches Innovationsmanagement: Erfolgsfaktoren - Grundmuster - Fallbeispiele. 4*, überarbeitete ed. Wiesbaden: Gabler Verlag/GWV Fachverlage GmbH.

Taylor, Frederick W. 1911. *The principles of scientific management.* New York: Harper & Brothers.

Tukker, Arnold. 2004. Eight types of product–service system: Eight ways to sustainability? Experiences from SusProNet. *Business Strategy and the Environment* 13(4): 246–260. doi:10. 1002/bse.414.

van Marrewijk, Marcel. 2003. Concepts and definitions of CSR and corporate sustainability: Between agency and communion. *Journal of Business Ethics* 44(2–3): 95–105. doi:10.1023/A: 1023331212247.

van Marrewijk, Marcel, and M. Werre. 2003. Multiple levels of corporate sustainability. *Journal of Business Ethics* 44(2): 107–119. doi:10.1023/A:1023383229086.

VDI. 2016. *VDI-Richtlinie: VDI 4070 Blatt 1 Nachhaltiges Wirtschaften in kleinen und mittelständischen Unternehmen - Anleitung zum nachhaltigen Wirtschaften*, no. VDI 4070. Berlin: Beuth.

Vernadat, François B. 1996. *Enterprise modeling and integration: Principles and applications*, 1. ed. London u.a. Chapman & Hall.

Welge, Martin K. 2001. *Strategisches Management: Grundlagen - Prozess - Implementierung*, 3rd ed. Wiesbaden: Gabler.

Westkämper, Engelbert, and Markus Decker. 2006. *Einführung in die Organisation der Produktion.* Berlin, Heidelberg: Springer.

Wilding, Richard, Joe Miemczyk, Thomas E. Johnsen, and Monica Macquet. 2012. Sustainable purchasing and supply management: A structured literature review of definitions and measures at the dyad, chain and network levels. *Supply Chain Management: An International Journal* 17 (5): 478–496. doi:10.1108/13598541211258564.

Will, Markus. 2012. *Strategische Unternehmensentwicklung auf Basis immaterieller Werte in KMU: Eine Methode zur Integration der ressourcen- und marktbasierten Perspektive im Strategieprozess.* ed. Kai Mertins. Berichte aus dem Produktionstechnischen Zentrum Berlin. Stuttgart: Fraunhofer.

Windolph, Sarah E., Dorli Harms, and Stefan Schaltegger. 2014. Motivations for corporate sustainability management: Contrasting survey results and implementation. *Corporate Social Responsibility and Environmental Management* 21(5): 272–285. doi:10.1002/csr.1337.

Wirtz, Bernd W. 2010. *Business model management: Design—Instrumente - Erfolgsfaktoren von Geschäftsmodellen*, 1st ed. Wiesbaden: Gabler.

Yunus, Muhammad, Bertrand Moingeon, and Laurence Lehmann-Ortega. 2010. Building social business models: Lessons from the Grameen experience. *Long Range Planning* 43(2–3): 308–325. doi:10.1016/j.lrp.2009.12.005.

Part V
Implementation Perspectives

Sustainable Value Creation—From Concept Towards Implementation

Steve Evans, Lloyd Fernando and Miying Yang

Abstract Sustainability is crucial to create long-term high value in manufacturing system. Sustainable value creation requires systems thinking in order to maximise total value captured. There is a need to better understand how companies can improve sustainable value creation. Few tools or structured approaches to thinking about sustainable value are available. This chapter seeks to provide understanding of key concepts for and tools that aid practitioners in sustainable value creation in manufacturing. The chapter also provides case studies on how the tools have helped companies improve sustainability.

Keywords Sustainable value creation · System thinking · Cambridge Value Mapping Tool · Sustainable Value Analysis Tool · Business model innovation · Sustainable business models

1 Introduction

We currently live in a world of constrained resources, growing populations and exceeding planetary boundaries. There is a need for industry to change the way we make things and shift towards a more sustainable industrial system. Understanding of system transformation and value transformation are important concepts for transitioning towards a more sustainable industrial system. Senge (1990) states that the un-healthiness of the world today is indirect proportion to our inability to see it

S. Evans (✉)
Centre for Industrial Sustainability, University of Cambridge, Cambridge, UK
e-mail: se321@cam.ac.uk

L. Fernando
Expo Industrial Engineering, University of Cambridge, Cambridge, UK

M. Yang
University of Cambridge, Cambridge, UK

© The Author(s) 2017
R. Stark et al. (eds.), *Sustainable Manufacturing*, Sustainable Production,
Life Cycle Engineering and Management, DOI 10.1007/978-3-319-48514-0_13

as a whole. Companies may not be fully aware of the full range of potential value outcomes. Most existing business models are mostly based on creating, delivering and capturing economic value, with limited or no attention to environmental and social value. The changing business environment, wider range of stakeholders engaging in debate over industry, resource limitations and emphasis on social responsibilities of firms has raised the need for sustainable value creation.

2 Key Concepts for Sustainable Value Creation

The industrial sustainability literature reviewed suggests system thinking and whole system design techniques as being one of the critical ways to understand sustainable value. This section presents main ideas on system thinking, whole system design, systems innovation and sustainable business models as the key concepts for sustainable value creation.

2.1 Systems Thinking

Seiffert and Loch (2005) suggest that the most important property of systems is that they are made up of several parts that are not isolated, but closely interlinked, forming a complex structure. Systemic or systems thinking, facilitates the improved understanding of these complex systems and enables the identification and utilisation of interrelationships and linkages as opposed to things.

Systems thinking is a technique for investigating entire systems, seeking to understand the relationships, the interactions, and the boundaries between parts of a system (Senge et al. 2008; Cabrera and Cabrera 2015). Systems thinking is particularly well suited to modeling highly complex open-systems where an integrated understanding is required at both the micro and macro-levels in order to predict or manage change. This contrasts with the dominant analytical approach of the physical sciences, which is based on reductionism, analysing closed-systems at the level of their constituent parts and then simplifying to draw out general conclusions. Systems thinking is a generic term that spans a range of more than 20 tools and methodologies (Reynolds and Holwell 2010).

Senge (1990) explains that systems thinking is a discipline for seeing wholes. It is a framework for seeing interrelationships rather than things, for seeing patterns of change rather than static snapshots. It appears that systems thinking is a way of approaching problems: rather than applying a strict linear methodology, the techniques are iterative, and designed to stimulate investigation, discussion and debate by encouraging multiple perspectives. Systems thinking does not aim to provide quantifiable answers to specific problems, but rather provides a range of options and better understanding of the implications of those options (Meadows and Wright

2009; Madrazo and Senge 2011). Manzini and Vezzoli (2003) emphasise the need for design for sustainability to move from product thinking to system thinking.

Network analysis potentially provides the scope to integrate multiple factors (economic, social and environmental). Preliminary research on analysing sustainability within industrial networks has demonstrated the use of such tools in understanding how and why networks adopt sustainability initiatives and the significance of 'focal' companies within the network (Van Bommel 2011).

It is described by authors (e.g. Senge et al. 2008) that many of the current challenges in industrial systems stem from the inability to understand and manage dynamic systems. Systems Thinking takes a birds-eye view and observes the whole picture by focusing on the relationships between the different entities of a system, rather than on isolated parts. Systems thinking is described by authors (Hawken et al. 1999; Rocky Mountain Institute 2006; Senge et al. 2008; Evans et al. 2009; Charnley et al. 2011; Cabrera and Cabrera 2015) as providing the foundation for a proactive approach to be able to design sustainable industrial systems (e.g. Systems Thinking can be a way to understand complex, non-linear, and interconnected systems of businesses, whether social, managerial, economical or environmental issues). There is lack of evidence and understanding of what abilities do companies need to improve their industrial sustainability at systems level. An ability-based view is not presented.

2.2 Whole System Design

Whole systems design is one approach to sustainable design offering great potential, however the processes, principles, and methods guiding the whole systems approach are not clearly defined or understood by practicing designers or design educators (Charnley et al. 2011).

Evans et al. (2009) describes whilst it is important to address the impact of each aspect of the industrial system and pursue aggressive reduction in the impact of specific activities, we must also examine the operation of the whole system. Efficiently manufacturing products that are inefficient in use, for example, is not enough. This approach can even result in substantially negative outcomes when efficiency gains or cost reductions result in increases in consumption (the so-called Rebound Effect). The greatest opportunity to reduce the impact of the industrial system on the planet arises when we consider the whole system and the optimisation of any individual component of the industrial system.

Rocky Mountain Institute-RMI (2006) define whole system design as 'optimising not just parts but the entire system ... it takes ingenuity, intuition, and teamwork. Everything must be considered simultaneously and analysed to reveal mutually advantageous interactions (synergies) as well as undesirable ones'. Whole-systems thinkers see wholes instead of parts, interrelationships and patterns, rather than individual things and static snapshots. They seek solutions that simultaneously address multiple problems (Anarow et al. 2003). Lovins (2011) are among the small number of authors who suggest that understanding the dynamics of a system is integral to the whole system approach. The Rocky Mountain Institute (2004) highlights systems thinking as the method that should be utilised not only to point the way to solutions to particular resource problems, but also to reveal interconnections between problems, which often permits one solution to be leveraged to create many more. Meadows (2009) lists nine places to intervene in a system, in increasing order of impact: numbers (subsidies, taxes, standards), material stocks and flows, regulating negative feedback loops, driving positive feedback loops, information flows, the rules of the system (incentives, punishment, constraints), the power of self-organisation, the goals of the system, and the mindset or paradigm out of which the goals, rules, and feedback structures arise.

It is suggested by the authors that reframing the system with a whole systems view helps people to understand more fully the way manufacturing affects the world we live in and how we might begin to change it (i.e. redesign the industrial system). Understanding who is involved in the current system and how they interact with it can help identify more opportunities to create sustainable value. The field of whole systems design and the literature surrounding it remains limited (Coley and Lemon 2009). Evans et al. (2009) describes the evidence from the case studies implementing and shifting towards more sustainable manufacturing and demonstrates that dramatic improvements can be made at the level of sub-systems, such as factories or businesses. In parallel, however, it will be necessary to develop the understanding and capabilities necessary to enable changes in the whole industrial system. Anarow et al. (2003) state that "sustainability cannot be achieved in the absence of whole-systems thinking", an ability that appears to be essential to improve industrial sustainability performance.

2.3   Systems Innovation

It is argued the innovations required for sustainable development need to move beyond incremental adjustments. Sustainable development requires the transformation of larger parts of production and consumption systems (Boons 2009). Incremental (product- and process-related) innovations in existing production and consumption systems may lead to further gradual improvements of sustainability performance, but in the end, incremental innovation frequently does not lead to a globally optimal system configuration in a multi-dimensional production and consumption system space (Larson 2000; Frenken et al. 2007; Vezzoli et al. 2008; Schaltegger and Wagner 2011).

While the term sustainable innovation has been widely used during the last decade, the number of definitions in the academic literature is limited (Holmes and Smart 2009; Boons and Lüdeke-Freund 2013). The review by Carrillo-Hermosilla et al. (2010) lists innovation definitions that focus on ecological sustainability, such as eco-innovation and environmental innovation. For instance, Carrillo-Hermosilla et al. (2010) introduced their own definition of eco-innovation: "innovation that improves environmental performance". Charter et al. (2008) describes that given the challenges posed by sustainable development, sustainable innovation will often be characterised by systemness and radicalness. Generally, sustainable innovations go beyond regular product and process innovations and are future-oriented. Sustainable innovation goes beyond eco-innovation because it includes social objectives and is more clearly linked to the holistic and long-term process of sustainable development for the short- and long-term objectives of sustainability. Holmes and Smart (2009) describe the need for more research in sustainability-led innovations and partnerships.

Adams et al. (2016) presents a model of (SOI) sustainability-oriented innovation onto which sustainability oriented innovation practices and processes can be mapped:

- Operational optimisation (e.g. eco-efficiency—compliance, efficiency, doing the same things better)
- Organisationtinal transformation (e.g. new market opportunities—novel products, services or business models, doing good by doing new things)
- Systems building (e.g. societal change—novel products, services or business models that are impossible to achieve alone, doing good by doing new things with others).

Adams et al. (2016) describe sustainability-oriented innovation as making intentional changes to an organisation's philosophy and values, as well as to its products, processes or practices to serve the specific purpose of creating and realising social and environmental value in addition to economic returns.

Draper (2015) in the report—'Creating the big shift: system innovation for sustainability, defines systems innovation as "a set of actions that shift a system—a city, a sector, an economy—onto a more sustainable path". It is described in this definition; being able to identify the set of actions is important, systems change usually requires multiple interventions across different areas of society, it is very rare that a single person or innovation can change a whole complex system, such as waste or energy and tackling problems that are too large for any one organisation, however powerful, to solve on its own (e.g. shift systems to make them more resilient, more equitable and able to continue into the future). Draper (2015) states that there is an "absence of necessary skills in sectors that can take the innovation to scale".

Sustainable development is argued by some authors to require radical and systemic innovations. Some authors argue these innovations can be more effectively created when building on the concept of business models. Sustainable business models provide the conceptual link between sustainable innovation and economic performance at higher system levels (Boons and Lüdeke-Freund 2013). Sustainable innovation is described by some authors to often be characterised by radicalness, some argue sustainable innovations go beyond regular product and process innovations and are future-oriented (Charter et al. 2008). Sustainable innovation is described by Charter et al. (2008) "Sustainable innovation is a process where sustainability considerations (environmental, social, and financial) are integrated into company systems from idea generation through to research and development (R&D) and commercialisation. This applies to products, services and technologies, as well as to new business and organisational models".

2.4 Sustainable Business Models

Bocken et al. (2014) states that business model innovations for sustainability are defined as: innovations that create significant positive and/or significantly reduced negative impacts for the environment and/or society, through changes in the way the organisation and its value-network create, deliver value and capture value (i.e. create economic value) or change their value propositions. It is argued in Bocken et al. (2014) that to tackle the pressing challenges of a sustainable future, innovations need to introduce change at the core of the business model to tackle unsustainability at its source rather than as an add-on to counter-act negative outcomes of business. The level of ambition of business model innovations needs to be high and focused on maximising societal and environmental benefits, rather than economic gain only. The sustainable business model innovation describing radical changes in the way companies do business has received considerable attention from both academia and practitioners (Chesbrough 2010; Zott et al. 2011). Sustainability management deals with social, environmental and economic issues in an integrated manner to transform organisations in a way that they contribute to a sustainable development of the economy and society within the limits of the ecosystem. Leaders, managers and entrepreneurs are challenged to contribute to sustainable development on the individual, organisational and societal level. Scholars and practitioners are recently increasingly exploring if and how modified and completely new business models can help maintain or even increase economic prosperity by either radically reducing negative or creating positive external effects for the natural environment and society, literature surrounding this area is scarce and still emerging.

Organisations today are challenged to contribute to sustainable development on the individual, organisational and societal level. Sustainability management refers to approaches dealing with social, environmental and economic issues in an integrated manner to transform organisations in a way that they contribute to a sustainable development of the economy and society within the limits of the ecosystem e.g. (Starik and Kanashiro 2013; Schaltegger et al. 2012; Boons and Lüdeke-Freund 2013). It appears "technological fix"—is insufficient to create the required transformation of organisations, industries and societies towards more sustainability. Researchers and practitioners are therefore increasingly exploring how completely new business models can help maintain or even increase economic prosperity by either radically reducing negative or creating positive external effects for the natural environment and society e.g. (Boons and Lüdeke-Freund 2013; Hansen et al. 2009; Schaltegger et al. 2012; Stubbs and Cocklin 2008). This perspective does not only cover existing organisations and how their business models are transformed (e.g. Sommer 2012), but also entirely new business models pioneered by entrepreneurs. The literature on sustainable business models is still emerging.

The literature presents numerous views on what constitutes a business model (e.g. Richardson 2008). Teece (2010) provides a concise definition: a business model is the design or architecture of the value creation, delivery and capture mechanism of a firm, how the firm delivers value, how it attracts customers, and how it converts this to profit (Teece 2010). Richardson proposes a summary organised around the concept of value:

- The value proposition—offering, target customer, differentiation;
- The value creation and delivery system—The value chain required, resources, assets, processes, position in the value network relative to customers, competitors and collaborators;
- The value capture system—How the firm makes money (financial model) and competitive strategy.

Evans et al. (2015) describe manufacturers are increasingly experimenting with new ways of meeting customers' needs. This includes shifting from providing products to providing services, in a way that separates the use of a product from its ownership; or circular economy models where products are designed and manufactured for continuous reuse, and value is captured from 'waste' wherever possible.

The sustainable business model literature describes the concept of value proposition and the creation of creative positive benefits to its stakeholders. There a growing volume of industrial cases on sustainable business models, but little is known on how these improvements were conceived, little is available about specific abilities and competencies (Barth et al. 2007; Segalas et al. 2009; Willard et al. 2010; Teece 2010; Bocken et al. 2014). System transformation and value transformation appear to be importance concepts to the research enquiry.

2.5 New Concepts for Sustainable Value Creation—Negative Forms of Value

Very few authors have contributed towards understanding the creation of new systems and generating value across the value network in the sustainable business models literature by identifying failed value exchanges. Authors such as (Rana et al. 2013; Yang et al. 2013; Bocken et al. 2014) are the few authors that have contributed

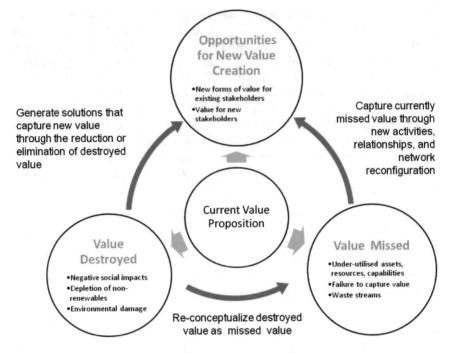

Fig. 1 Value propositions (Rana et al. 2013)

towards understanding opportunities for value creation. Yang et al. (2014) describe and define multiple forms of value (e.g. value absence, value surplus, value destroyed, value missed). Rana et al. (2013) and Bocken et al. (2014) in their research propose a framework for business model innovation for sustainability by explicitly considering value destroyed and value missed within the business model, as these often represent important opportunities for sustainability innovation. Their research provides a qualitative framework to facilitate systematic exploration of the different forms of value for each stakeholder (Fig. 1).

- Value captured—current value proposition
- Value destroyed—negative value outcomes of current model
- Value missed—value currently squandered, lost or inadequately captured by current model
- Value opportunities—new opportunities for additional value creation and capture through new activities and relationships.

Based on this, Yang et al. (2016) further propose value uncaptured as a new perspective for sustainable business model innovation. Value uncaptured is defined

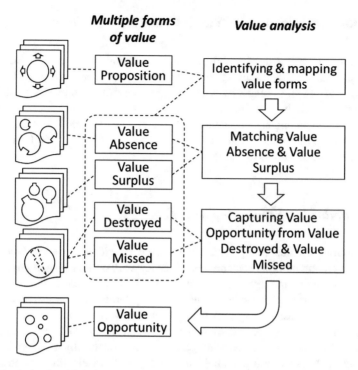

Fig. 2 Analysis of multiple forms of value (Yang et al. 2013)

as the potential value, which could be captured but has not been captured yet. Four forms of value uncaptured, i.e. value surplus, value absence, value destroyed and values missed and an approach of analysis of multiple forms of value was proposed shown in Fig. 2 (Yang et al. 2013).

Value uncaptured exists in almost all companies. Some uncaptured value is visible, e.g. waste streams in production, co-products, under-utilised resources, and reusable components of broken products; some is invisible, e.g. over capacity of labour, insufficient use of expertise and knowledge. Reducing any kind of the uncaptured value would create sustainable value. Yang et al. (2016) propose a framework of using value uncaptured for sustainable business model innovation, and claims that sustainable business model innovation can be more easily achieved by identifying the value uncaptured in current business models, and then turning this new understanding of the current business into value opportunities that can lead to new business models with higher sustainable value.

3 Tools for Sustainable Value Creation

This section describes the Cambridge Value Mapping Tool, and the Sustainable Value Analysis Tool and their strengths and weakness. The tools provide a structured way of helping companies identify opportunities for business model innovations that result in more sustainable businesses. This could assist companies maximise value among stakeholders across the system. The tools also provide new perspectives on sustainable value creation and aid transforming the businesses to deliver uncaptured and sustainable value.

3.1 Cambridge Value Mapping Tool

The Cambridge Value Mapping Tool has been developed to elicit failed value exchanges among multiple stakeholders in the network of the firm and uncover new value opportunities through a structured and visual approach. It is developed to assist manufacturing companies in identifying opportunities for sustainable value creation. The tool assists in systematically analysing various forms of value in your business and your network and stimulate innovation in sustainable value creation. The tool adopts a multi-stakeholder perspective, through which the exchange of value can be analysed and potential stakeholder conflicts identified to create positive value in the network. It provides a new perspective for practitioners to understand and create new economic, social, and environmental value from their business. The tool gives practitioners a new way to gain a deeper understanding of value and create new economic, social, and environmental benefits for their business (Fig. 3).

The Cambridge Value Mapping Tool was developed at the IfM's Centre for Industrial Sustainability by a research team led by Professor Steve Evans. Originating from the EU FP7 Sustain Value project, the tool since has gone through multiple conceptual and visual iterations. Acknowledgements for their contributions go to Dr. Padmakshi Rana, Dr. Samuel Short, Dr. Nancy Bocken, Dr. Dai Morgan, Dr. Miying Yang, Dr. Lloyd Fernando, Dr. Doroteya Vladimirova, Dr. Curie Park, Fenna Blomsma and Dr. Maria Holgado. Particular thanks to all industry collaborators who took part in the development, testing and refinement of the tool.

The Cambridge Value Mapping Tool takes you in a guided step-by-step process through the following questions:

- What is the unit of analysis e.g. product, service, company, industry?
- Who are the stakeholders for the unit of analysis?
- What is the purpose of the unit of analysis?
- What is the current value captured?

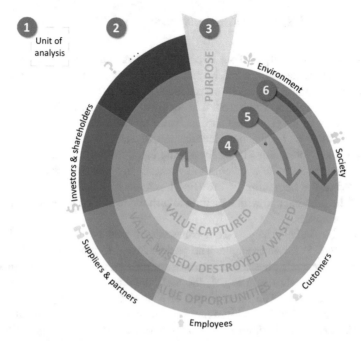

Fig. 3 Cambridge Value Mapping Tool (*Source* http://www.ifm.eng.cam.ac.uk/news/the-cambridge-value-mapping-tool/#.V8aiy5N961s)

- What is the value missed and/or destroyed?
- What is the value surplus and/or absence?
- What are the new value opportunities?

Strengths

- The tool can be used by individuals to identify opportunities to create sustainable value in their own companies.
- The tool gives practitioners a new way to gain a deeper understanding of value and create new economic, social, and environmental benefits for their business
- Designed to stimulate innovation of the business model for sustainable value
- Helps practitioners to find and create new economic, social, and environmental value from their business through a systematic analysis of various forms of value in the business and the firm's network
- Provides a structured approach to identify sustainable value opportunities

Weakness

- Does not explore the unintended consequences that can arise in other parts of the system for implementing the identified value opportunity.

3.2 *Sustainable Value Analysis Tool (SVAT)*

Sustainable Value Analysis Tool is built to help manufacturers identify opportunities to create sustainable value by analysing the captured and uncaptured value throughout the entire life cycle of products (Yang 2015). Identifying the value uncaptured and creating value from it is not always easy. The rationale of the tool is to use separate forms (i.e. value surplus, value absence, value destroyed and value missed) of value to inspire the identification of value uncaptured, and to further identify value opportunities by analysing the identified value uncaptured. The tool provides companies with a scheme to systematically look for each form of value uncaptured at the beginning of life (BoL), middle of life (MoL) and end of life (EoL) of the product, and with a method to turn the identified value uncaptured into value opportunities.

Sustainable Value Analysis Tool consists of a poster (see Fig. 4) and a set of cards (see Fig. 5) for an example. The poster is used for gathering insights across the different life cycle phases and the cards for guiding and inspiring the process of using the tool. As shown in Fig. 4, the tool combines the life cycle thinking and value forms analysis. The three phases of a product life cycle (BOL, MOL and EOL) could be further divided into more specific stages. For example, MOL can be further divided into distribution, use, maintenance and service. The value forms

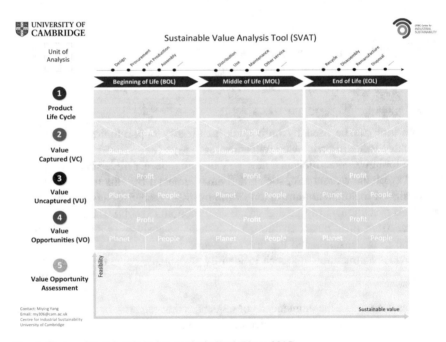

Fig. 4 Poster of Sustainable Value Analysis Tool (Yang 2015)

Fig. 5 Cards of Sustainable Value Analysis Tool (Yang 2015)

consist of value captured, value uncaptured and value opportunities. Value uncaptured could be considered from the perspectives of value destroyed, value missed, value surplus and value absence.

Sustainable Value Analysis Tool mainly consists of five steps:

- Step 1. Define the life cycle stages of a product in the company, and map the stakeholders involved in each stage of product life cycle
- Step 2. Describe what is the value captured for each stakeholder (environmental, social and economic dimensions) in each stage of the defined product life cycle
- Step 3. Identify what is the value uncaptured for each stakeholder (environmental, social and economic dimensions) in each stage of the defined product life cycle
- Step 4. Identify value opportunities, e.g. how to turn value uncaptured into value opportunities
- Step 5. Assess the feasibility and sustainability of each identified value opportunity

For each step there is a card providing step-by-step guidance including background knowledge, tasks and tips on the front and some inspirational examples on the back.

The tool can elicit value uncaptured across products life cycle, and uncover new value opportunities through a structured and visual approach.

Strengths

- Comprehensive analysis of value
- Generating business opportunities in a strategic way (by turning value uncaptured into value opportunities)

- Innovation for sustainability
- Embedding stakeholder theory and life cycle thinking
- Business model driven

  Weakness

- Does not include strategic planning on how to realise the identified opportunities.

4   Case Studies: Lessons Learnt from Practice on Sustainable Value Creation

This section elaborates on the cases investigated to explore the current industrial practice in business models and identify failed value exchanged and find opportunities to capture value. For confidentiality purpose the names of the firm and the interviewees have not been revealed.

Introduction

Company A is a fast moving consumer good, Sugar manufacturer. The case studies of this company provide a generic view of value exchanges between firm and stakeholder groups.

Company A aims to transform all raw materials into sustainable products. The plant in Wissington has been operating for over 85 years and now produces over 420 kt of sugar annually for food and drinks manufacturers The company uses a culture of innovation to reduce process inputs, minimise waste and deliver its commitment to be an advanced and sustainable manufacturer. The company has been able to find ways of internalising and being very effective at it. The company converts raw beet to sugar and the byproducts are used to produce electricity, tomatoes, animal feed, and other materials. No material arriving into the company is allowed to disappear as waste (and a cost). Instead all materials are turned into valuable co-products, including the soil attached to the beet, which becomes clean soil for gardeners, these actions contribute to a very high level of efficient use of raw materials. The   company has been able to bring more value under its control and link knowledge to benefit by turning everything into a valuable output.

Data

We are the world's largest refinery producing 420,000 tonnes of Sugar annually… We been able to find opportunities in our process to produce co-products from the waste streams of the primary sugar production processes… (Symbiotic co-product lines)… We have found a broad range of additional synergistic and profitable product lines… animal feed, electricity, tomatoes, and bioethanol… More than two hundred and forty miles of piping carries hot water from the factory's Combined Heat and Power (CHP) plant around the glasshouse, to maintain the balmy temperatures, which suit tomato plants. This hot water would otherwise be destined for cooling towers, so the scheme ensures that the heat is used productively…. carbon dioxide as a by-product from the CHP boiler is pumped into the enormous glasshouse

to be absorbed by the plants (rather than vented into the atmosphere as waste emissions)...
waste carbon dioxide from the factory is used by tomatoes for photosynthesis... the site also
harvests the rainwater from the giant glasshouse roof; over 115 million litres are collected
annually to irrigate the plants...the horticulture business produces around 140 million
'eco-friendly' tomatoes each year...co-product generated by finding opportunities for productive, and creative use of the waste streams....The heated atmosphere of 4 times ambient
levels of CO2 enables the tomatoes to grow at twice the usual rate, providing high productivity for the glasshouse investment (Interviewee 2B—Head of Engineering).

Analysis—From Concept Towards Implementation

The data suggests the company for example a leader in efficiently and sustainably
manufacturing sugar beet, over the past three decades has been able to systematically find failed value exchanges in their system. The company described, "We
routinely seek innovative ways to minimise waste and maximise value". The
company has been able to see 'carbon emissions' and 'low-grade heat' escaping
from its processes into the atmosphere as a failed value (a by-product from the CHP
boiler). The company described, "this hot water would otherwise be destined for
cooling towers... we identified that our supply of carbon dioxide, heat and water
could be better exploited if we used it again." The company has been able to
identify the waste streams (i.e. carbon dioxide, heat) that had value that is not being
captured and destroyed in its system (i.e. failed value).

The data suggests that company for example has been able to turn waste streams
(i.e. failed value) and emissions from their core production processes into useful
and positive inputs to new product lines. No material arriving into the company is
allowed to disappear as waste (and a cost). Instead all materials are turned into
valuable co-products. The data suggests that the company has been able to firstly
identify failed values and then bring more value under its control by using and
linking its knowledge to turn waste streams in its current systems into a valuable
output and create positive value. The company has been able to see the combustion
gases from the power station and low-grade heat as failed value lost to the atmosphere. The company described how it has been able to find away to capture the two
waste streams and transform it to create new positive value (i.e. grow tomatoes) and
deliberately bring it into the business model. By seeing failed value and bringing it
into the business model, the company has been able to make productive use of
waste carbon dioxide and heat from the sugar factory, which tomatoes (new
co-product) use during photosynthesis. It is described the carbon dioxide (a
by-product from the CHP boiler) is pumped into the enormous glasshouse to be
absorbed by the plants, rather than vented into the atmosphere as waste emissions. It
is observed the company has firstly been able to see the failed value exchange, and
then figure out what to do with it to form positive value, and come up with a
solution using its knowledge and control.

5 Conclusion

This chapter provides key concepts for increasing sustainable value creation in manufacturing, and presents the tools which can help companies using the concepts in practice. Sustainable value creation requires companies to have systems thinking when making business decisions. Companies need to consider the value creation for multi-stakeholders, including customers, suppliers, employees, society and planet. The concept of failed value exchange is identified to be helpful for companies to identify opportunities for sustainable value creation. The evidence suggests that by looking at what value exchanges are failing across the multiple stakeholders, organisations are found to be able to see a lot of value opportunities. The system transformation that industry needs requires more cross-business system collaboration. A case study of sugar manufacturer is provided to illustrate how these concepts are implemented in industries.

References

Adams, R., S. Jeanrenaud, J. Bessant, D. Denyer, and P. Overy. 2016. Sustainability-oriented innovation: A systematic review. *International Journal of Management Reviews* 18: 180–205.

Anarow, B., C. Greener, V. Gupta, M. Kinsley, J. Henderson, C. Page, and K. Parrot. 2003. Whole-systems framework for sustainable consumption and production. *Report for Danish Ministry of the Environment, Denmark* 807: 1–51.

Barth, M., J. Godemann, M. Rieckman, and U. Stoltenberg. 2007. Developing key competences for sustainable development in higher education. *International Journal of Sustainable Higher Education* 8: 416–430.

Bocken, N., S.W. Short, and S. Evans. 2014. A literature and practice review to develop sustainable business model archetypes. *Journal of Cleaner Production* 65: 42–56.

Boons, F. 2009. *Creating ecological value. An evolutionary approach to business strategies and the natural environment*. Cheltenham: Edward Elgar.

Boons, F., and F. Lüdeke-Freund. 2013. Business models for sustainable innovation: state-of-the-art and steps towards a research agenda. *Journal of Cleaner Production* 45: 9–19.

Cabrera, D., and L. Cabrera. 2015. *Systems thinking made simple: New hope for solving wicked problems*. Ithaca, NY: Odyssean.

Carrillo-Hermosilla, J., P. del Río, and T. Könnölä. 2010. Diversity of eco-innovations: Reflections from selected case studies. *Journal of Cleaner Production* 18: 1073–1083.

Charnley, F., M. Lemon, and S. Evans. 2011. Exploring the process of whole system design. *Design Studies* 32(2): 156–179.

Charter, M., C. Gray, T. Clark, and T. Woolman. 2008. Review: The role of business in realising sustainable consumption and production. *System Innovation for Sustainability: Perspectives on Radical Changes to Sustainable Consumption and Production* 46–69.

Chesbrough, H.W. 2010. Business model innovation: Opportunities and barriers. *Long Range Planning* 43(2): 354–363.

Coley, F., and M. Lemon. 2009. Exploring the uncertainty surrounding the design and perceived benefit of sustainable solutions. *Journal of Engineering Design* 20: 543–554.

Draper, S. 2015. Creating the big shift: System innovation for sustainability. The Forum for the Future, London. http://www.forumforthefuture.org. Accessed 9 October 15.

Evans, S., M.N. Bergendahl, M. Gregory, and C. Ryan. 2009. Towards a sustainable industrial system. In *International Manufacturing Professors' Symposium in Cambridge, UK.*

Evans, S., T. Moore, and M. Folkerson. 2015. *Industrial evolution: Making British manufacturing sustainable.* CAN Mezzanine, UK: Policy Connect.

Frenken, K., M. Schwoon, F. Alkemade, and M. Hekkert. 2007. A complex systems methodology to transition management. In *DRUID Summer Conference 2007, Copenhagen, CBS,* 18–20.

Hansen, E.G., F. Grobe-Dunker, and R. Reichwald. 2009. Sustainability innovation cube. A framework to evaluate sustainability-oriented innovations. *International Journal of Innovation Management* 13(4): 683–713.

Hawken, P., A.B. Lovins, and L.H. Lovins. 1999. *Natural capitalism: Creating the next industrial revolution.* Little, Brown and Co: Array Boston.

Holmes, S., and P. Smart. 2009. Exploring open innovation practice in firm-nonprofit engagements: A corporate social responsibility perspective. *R&D Management* 39(4): 395–409.

Larson, A.L. 2000. Sustainable innovation through an entrepreneurship lens. *Business Strategy and the Environment* 9: 304–317.

Lovins, A.B. 2011. *Reinventing fire: Bold business solutions for the new energy era.* White River Junction, VT: Chelsea Green Publishing.

Madrazo, C., and P. Senge. 2011. *Building communities of collaboration and co-inspiration.* Academy for Systemic Change.

Manzini, E., and C. Vezzoli. 2003. A strategic design approach to develop sustainable product service systems: Examples taken from the environmentally friendly innovation Italian prize. *Journal of Cleaner Production* 11(8): 851–857.

Meadows, D.H., and D. Wright. 2009. *Thinking in systems: A primer,* 1st ed. London: Earthscan Ltd.

Rana, P., S.W. Short, N. Bocken, and S. Evans. 2013. Towards a sustainable business form: A business modelling process and tools. In *Sustainable Consumption Research and Action Initiative (SCORAI) Conference: The Future of Consumerism and Well-Being in a World of Ecological Constraints, 12–14 June, Clark University, Worcester, USA.*

Reynolds, M., and S. Holwell. 2010. *Systems approaches to managing change: A practical guide,* 1st ed. London: Springer.

Richardson, J. 2008. The business model: an integrative framework for strategy execution. *Strategic Change* 17(5): 133–144.

Rocky Mountain Institute. 2004. Whole system design. http://www.rmi.org/. Accessed 20 August 2015.

Rocky Mountain Institute. 2006. Hypercar design and technology. https://old.rmi.org. Accessed 13 October 2015.

Schaltegger, S., F. Lüdeke, and E. Hansen. 2012. Business cases for sustainability: The role of business model innovation for corporate sustainability. *International Journal Innovation Sustainable Development* 6(2): 95–119.

Schaltegger, S., and M. Wagner. 2011. Sustainable entrepreneurship and sustainability innovation: Categories and interactions. *Business Strategy and the Environment* 20(4): 222–237.

Segalas, J., D. Ferrer-Balas, M. Svanstrom, U. Lundqvist, and K.F. Mulder. 2009. What has to be learnt for sustainability? A comparison of bachelor engineering education competencies at three European Universities. *Sustainability Science* 4(1): 17–27.

Seiffert, M., and C. Loch. 2005. Systemic thinking in environmental management: Support for sustainable development. *Journal of Cleaner Production* 13(12): 1197–1202.

Senge, P. 1990. *The fifth discipline: Mastering the five practices of the learning organization.* New York: Doubleday.

Senge, P.M., B. Smith, and N. Kruschwitz. 2008. *The necessary revolution: How individuals and organisations are working together to create a sustainable world.* New York: Doubleday.

Sommar, A. 2012. *Managing green business model transformations.* Heidelberg: Springer.

Starik, M., and P. Kanashiro. 2013. Toward a theory of sustainability management: Uncovering and integrating the nearly obvious. *Organization & Environment* 26(1): 7–30.

Stubbs, W., and C. Cocklin. 2008. Conceptualizing a sustainability business model. *Organization & Environment* 21(2): 103–127.

Teece, D.J. 2010. Business models, business strategy and innovation. *Long Range Planning* 43(2): 172–194.

Van Bommel, H.W.M. 2011. A conceptual framework for analyzing sustainability strategies in industrial supply networks from an innovation perspective. *Journal of Cleaner Production* 19 (8): 895–904.

Willard, M., C. Wiedmeyer, R.W. Flint, J.S. Weedon, R. Woodward, I. Feldmand, and M. Edwards. 2010. *The sustainability professional: 2010 competency survey report.* International Society of Sustainability Professionals.

Yang, M., P. Rana, and S. Evans. 2013. Product service system (PSS) life cycle value analysis for sustainability. In *Proceedings of the 6th conference on Design and Manufacture for Sustainable Development (ISBN: 978-1-84919-709-0), Zhejiang University, Hangzhou, China.*

Yang, M., D. Vladimirova, P. Rana, and S. Evans. 2014. Sustainable Value Analysis Tool for value creation. *Asian Journal of Management Science and Applications* 1(4): 312–332.

Yang, M. 2015. *Sustainable value creation for product-service systems (PSS).* University of Cambridge.

Yang, M., S. Evans, D. Vladimirova, and P. Rana. 2016. Value uncaptured perspective for sustainable business model innovation. *Journal of Cleaner Production.* doi:10.1016/j.jclepro. 2016.07.102.

Zott, C., R. Amit, and L. Massa. 2011. The business model: Recent developments and future research. *Journal of Management* 37(4): 1019–1042.

Life Cycle Sustainability Assessment Approaches for Manufacturing

Ya-Ju Chang, Sabrina Neugebauer, Annekatrin Lehmann,
René Scheumann and Matthias Finkbeiner

Abstract Sustainability assessments considering the three dimensions environment, economy, and society are needed to evaluate manufacturing processes and products with regard to their sustainability performance. This chapter focuses on Life Cycle Sustainability Assessment (LCSA), which considers all three sustainability dimensions by combining the three methods Life Cycle Assessment (LCA), Life Cycle Costing (LCC), and Social Life Cycle Assessment (SLCA). Existing LCSA approaches as well as selected ongoing work are introduced, both regarding the individual approaches as well as the combined LCSA approach. This includes, for instance, the Tiered Approach. This approach facilitates the implementation of LCSA, for instance, within the manufacturing sector, by providing a category hierarchy and guiding practitioners through the various impact and cost categories proposed for the three methods. Furthermore, ongoing developments in LCC and SLCA are presented, such as the definition of first economic and social impact pathways (linking fair wage and level of education to social damage levels) for addressing the current challenges of missing impact pathways for economic and social aspects. In addition, the Sustainability Safeguard Star suggests a new scheme for addressing the inter-linkages between the three sustainability dimensions. These approaches foster the application and implementation of LCSA and thus contribute to developing sustainable processes and products.

Keywords Life Cycle Sustainability Assessment (LCSA) · Sustainability assessment · Tiered approach · Life Cycle Assessment (LCA) · Life Cycle Costing (LCC) · Social Life Cycle Assessment (SLCA)

Y.-J. Chang (✉) · S. Neugebauer · A. Lehmann · R. Scheumann · M. Finkbeiner
Department of Environmental Technology, Technische Universität Berlin, Berlin, Germany
e-mail: ya-ju.chang@tu-berlin.de

R. Stark et al. (eds.), *Sustainable Manufacturing*, Sustainable Production,
Life Cycle Engineering and Management, DOI 10.1007/978-3-319-48514-0_14

1   Introduction

Sustainability and sustainable manufacturing are relevant topics for governments and industries worldwide. In that pursuit, various concepts for sustainability exist and approaches for sustainability assessment have already been introduced. Nevertheless evaluating the sustainability performance at the product level remains a challenge. One of the most widespread concepts of sustainability lies in the triple-bottom-line theory, which considers environmental, economic and social aspects (Finkbeiner et al. 2010; Remmen et al. 2007; Elkington 1998). Moreover, with regard to assessing the sustainability performance of products and processes, life cycle thinking approaches which include the whole life cycle from "cradle to grave," are increasingly gaining in importance. By employing such approaches, a shifting of impact between the different life stages and sustainability dimensions can be identified and avoided (Finkbeiner et al. 2010).

By combining both the triple-bottom line theory and life cycle thinking approaches, the Life Cycle Sustainability Assessment (LCSA) framework has been proposed as a mean of evaluating the sustainability performance of products. LCSA analyses environmental, economic and social sustainability aspects by combining the methods Life Cycle Assessment (LCA), Life Cycle Costing (LCC), and Social Life Cycle Assessment (SLCA). The LCSA framework has been initiated with the development of the "Product Portfolio Analysis" (PROSA; German: Produktlinienanalyse) (Öko-Institut 1987; Rainer Grießhammer et al. 2007) and was further developed and formulated by Klöpffer and Finkbeiner (Klöpffer 2008; Finkbeiner et al. 2010). LCSA has so far been identified and promoted as a feasible framework for measuring the performance of products in the three sustainability dimensions (UNEP 2012; Valdivia et al. 2012).

Yet, challenges in LCSA's applicability, scientific robustness, comprehensiveness, interpretation and practical implementation persist (Valdivia et al. 2012; Lehmann 2013; Neugebauer et al. 2015). These challenges mainly relate to the different maturity levels of the three methods considered. LCA is widely accepted and used in practice for assessing a variety of products and services (including e.g. technologies). Although LCA still contains some challenges (Finkbeiner et al. 2014), its general application and implementation stand unhindered. Yet, to date, SLCA and LCC have not yet reached a mature level of assessment. Their main methodological difficulties lie in insufficient guidance on indicator selection, missing sets of defined impact categories and areas of protection (AoPs, also called safeguard subjects), as well as missing links between indicators, impacts and AoPs (Valdivia et al. 2012; Lehmann 2013; Neugebauer et al. 2015, 2016). To overcome these challenges, new approaches have been proposed. One of them is the Tiered Approach, which provides a category hierarchy to facilitate the implementation of LCSA, for instance, in the manufacturing sector. Furthermore, social impact pathways (e.g. fair wage) have been defined and a new LCC approach (the economic LCA framework) has been proposed, addressing some of the methodological challenges associated with LCSA.

The following subsections present the three underlying methods of LCSA in detail, including state-of-the-art, research needs and outlook, elaboration on the application of LCSA in manufacturing (e.g. by using the Tiered approach), followed by an introduction to selected developments for improving on the LCSA framework.

2 Life Cycle Sustainability Assessment (LCSA)

As aforementioned, the LCSA framework consists of the three methods Life Cycle Assessment (LCA), Life Cycle Costing (LCC), and Social Life Cycle Assessment (SLCA), and thus considers positive and negative environmental, social and economic impacts. This combination of different life cycle methods is illustrated by the following Eq. (1) (Klöpffer 2008), which provide helpful guidance in the decision-making processes towards more sustainable products (UNEP/SETAC Life Cycle Initiative 2011).

$$LCSA = LCA + LCC + SLCA \qquad (1)$$

In the following sections, the state-of-the-art of the three methods within LCSA as well as their contribution to sustainable manufacturing are introduced. In addition further research needs and outlook are described.

2.1 Life Cycle Assessment (LCA)

LCA analyses the potential environmental impacts of products and processes from a life cycle perspective. The current development of LCA, and the research needs and outlook are introduced in the following sections.

2.1.1 State-of-the-Art

According to the European Commission (2015), LCA is the best available tool for evaluating the potential environmental impacts of manufacturing processes or products from cradle-to-grave. LCA is an ISO-standardised (ISO 2006a, b) method and structured into four phases: (1) goal and scope definition, (2) life cycle inventory analysis, (3) life cycle impact assessment, and (4) interpretation. Based on the standardised phases, environmental impact can be assessed in an iterative process.

The relation between inventory results, midpoint and endpoint impact categories and AoPs is determined through impact pathways, as displayed in Fig. 1. Inventory

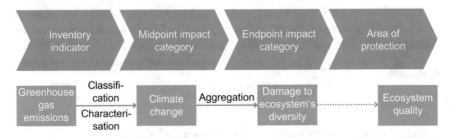

Fig. 1 Relation of inventory indicators, indicators on midpoint and endpoint impact category levels, and AoPs (exemplary illustration for greenhouse gas emissions)

indicators (e.g. greenhouse gas emissions) are classified into impact categories and characterised[1] at the midpoint level (e.g. climate change). The category indicator results achieved at the midpoint level can then be aggregated into impact categories at the endpoint level (e.g. damage to ecosystem's diversity). Those endpoint damage levels are then linked to AoPs (e.g. ecosystem quality).

After decades of method, database and software development, various case studies as well as international standardisation processes have emerged, so that one can now safely say that LCA has reached a mature stage and is robust enough to be applied in decision-making in both private organisations and governments (Finkbeiner et al. 2014).

2.1.2 Research Needs and Outlook

Although LCA has reached a mature level in implementation and has been internationally standardised, LCA still faces some challenges. Finkbeiner et al. (2014) identified 34 challenges with regard to inventory (e.g. dealing with allocation and delayed emissions), impact assessment (e.g. analysing impacts such as land use and odour), generic aspects (e.g. handling weighting and data quality analysis) and evolving aspects (e.g. considering littering, animal well-being or positive impacts), which have not been comprehensively addressed in the current literature and practice. Moreover, collecting relevant and robust data stands as an overall obstacle in carrying out LCA. Although several databases covering numerous different products and processes exist, specific applications (e.g. production of electronics) have so far been insufficiently contemplated. Work is currently ongoing to address some of the challenges, such as improving impact assessment methods (e.g. Bach and Finkbeiner 2016). Until challenges are resolved, practitioners should carefully

---

[1]The individual contribution of the emissions to the impact is calculated by multiplying the amount of each emission with a characterisation factor (for example, CH_4 has a 28 times higher contribution to global warming than CO_2).

check if the challenges identified limit the conclusions of LCA case studies (Finkbeiner et al. 2014).

2.2 Life Cycle Costing (LCC)

LCC evaluates different costs along the life cycle of a product or process in order to reflect the economic sustainability monetarily. Meanwhile, the current developments of LCC, the research needs in the context of LCSA, along with the overall outlook, are all introduced in the following sections.

2.2.1 State-of-the-Art

LCC appeared in the mid-1960s. Originally, it was used to rank different investment alternatives, but for a long time failed to consider operating costs occurring during the product's lifetime (Glucha and Baumann 2004). A first international standard was published in 2008 with ISO 15686-5 focusing on buildings and construction assets. Therein, LCC is defined as a tool that enables comparative cost assessments (in terms of initial costs and future operational costs) over a specified period of time (ISO 2008).

A similar approach was adopted by Hunkeler et al. (2008), who include producers, suppliers, consumers and end of life actors in the assessment for reflecting costs associated with a product's life cycle. They furthermore differentiate LCC into three types—conventional LCC, environmental LCC, and societal LCC. Conventional LCC focuses on internal costs directly associated with a product's life cycle. Environmental LCC goes beyond that scope and includes external costs likely to be internalised in the decision-relevant future, such as environmental taxes and subsidies (Hunkeler et al. 2008). Societal LCC even includes costs emerging from the side-effects of production which manifest in people's lives and society, whether today or in the long-term. Within the realm of LCSA, it is normally referred to as environmental LCC in the interest of avoiding overlap with the other two dimensions.

2.2.2 Research Needs and Outlook

Several challenges however hinder LCC's methodology development and thus implementation within the LCSA framework. They are, for example, oversimplifying the economic dimension down to a matter of costs, ignoring causalities, or unreliable data in connection with conceptual confusions (Neugebauer et al. 2016). To date, LCC in the context of LSCA is still not commonly implemented in industry, due to methodological confusion with other similar concepts, such as "total cost accounting" (Glucha and Baumann 2004). Furthermore, the limitation attached to costs has often been criticised especially in the context of LCSA. In contrast to LCA, LCC does not contain impact pathways following a cause-effect-chain. Consequently, several authors discuss whether LCC can sufficiently measure and represent economic sustainability within the LCSA framework (Jørgensen et al. 2010; Heijungs et al. 2013). The debate is associated with the question of whether or not LCC should stay at the cost level, or if the classical LCC framework should be extended to implement a broader economic perspective, e.g. by connecting costs on the microeconomic level to impact on the macroeconomic level. To mitigate this situation, May and Brennan (2006) suggested including value added (VA) as an economic indicator and relating it to wealth generation. Wood and Hertwich (2012) went even further by linking VA to gross domestic product through input-output modelling.

Furthermore, to bridge the gap in pursuit of aligning the economic dimension involved in LCSA with LCA, Neugebauer et al. (2016) proposed the concept of economic LCA (EcLCA), and defined midpoint and endpoint impact categories as well as AoPs for the economic dimension. This approach is further described in Sect. 4.1.2. Further research should focus on the definition of impact pathways as well as provision of concrete quantified measures for impact pathways.

2.3 Social Life Cycle Assessment (SLCA)

SLCA aims at analysing the social and socioeconomic impact of products and processes. In the following sections, the state-of-the-art, research needs and outlook for developing SLCA are presented.

2.3.1 State-of-the-Art

SLCA investigates the positive and negative social and socio-economic impact of products or processes along their life cycle. According to the 'Guidelines for SLCA of products' (UNEP/SETAC 2009), the impacts may affect the concerned stakeholder groups: workers, consumers, local communities, value chain actors and the society, and may be linked to the company's behaviour. Complying with the guidelines, the 'Methodological Sheets for Subcategories in SLCA' was published and provided practical guidance on the subcategories and potential indicators for conducting SLCA case studies (Benoît et al. 2013).

2.3.2 Research Needs and Outlook

Several deficiencies persist with the SLCA methodology and therefore impede its implementation in practice, e.g. in industry. Although the methodological sheets provided indicator sets related to relevant stakeholder groups, no widely agreed approach for selecting indicators, relevant social issues, and involved stakeholders exists (Lehmann et al. 2013; Martínez-Blanco et al. 2014; Andreas Jørgensen et al. 2009). In addition, since social impacts are usually associated with organisations' behaviour (Dreyer et al. 2006; Andreas Jørgensen et al. 2009), allocating social impact to a specific product is not straightforward and thus often hinders the implementation and meaningfulness of SLCA (Andreas Jørgensen 2013; Lehmann et al. 2013). Another big challenge lies in linking social indicators to impact categories and AoPs via social impact pathways (Lehmann et al. 2013; Neugebauer et al. 2014). Without such impact pathways, i.e. proper impact pathways and AoPs, a complete picture of potential social impacts cannot be fully anticipated. One of the first approaches for an impact pathway was developed by Jørgensen et al. (2010a, b), who developed impact pathways for child labour and also highlighted the difficulties in measuring the potential girth of the impact.

A more recent approach for impact pathways was provided by Neugebauer et al. (2014), proposing impact pathways for fair wage and the level of education. This approach is presented in more detail in Sect. 4.1.1. Further research is geared to focus on the development of databases and more impact pathways addressing social aspects beyond child labour, wage and education as well as regarding the concretisation of the impact pathways by providing e.g. concrete quantified impact pathways.

3 Application of LCSA in Manufacturing: Tiered Approach

So far, environmental indicators resulting from LCA or simplified LCA (e.g. carbon footprint) are widely employed in manufacturing sectors in order to evaluate the environmental performance of products or processes. Yet, economic and social indicators are currently just randomly considered in product or process assessments due to the methodological challenges associated with LCC and SLCA. Consequently, valid indicator sets for a holistic LCSA are currently lacking and thus hinder the implementation of LCSA in manufacturing sectors. A first attempt to foster application of LCSA is the Tiered Approach, which provides a step-by-step procedure going from a simplified LCSA to a comprehensive one (Neugebauer et al. 2015).

3.1 Framework of the Tiered Approach

The Tiered Approach is a "step-by-step" guidance for applying and implementing LCSA in practice (see Fig. 2). It provides an impact and cost categories hierarchy, which supports LCSA practitioners in selecting suitable indicators, and indicates potential directions of future development in LCSA. The categories proposed have been chosen from selected sources, e.g. the ILCD Handbook of LCA (JRC 2011), the Guidelines for SLCA of products (UNEP/SETAC 2009), and the Code of Practice for LCC (Swarr et al. 2011) based on three criteria (relevance, robustness of the methods, and practicality). For LCA, impact categories at midpoint level are selected since the midpoint results have more consensus characterisation methods and lower statistical uncertainty than the endpoint results (Bare et al. 2000).

Three tiers are recommended in the Tiered Approach: Tier 1, namely Sustainability Footprint, represents a "low entry-level" LCSA, where only few categories are considered (e.g. climate change, production costs and fair wages). Hence, Tier 1 provides a basis for aligning the different maturity levels of LCA, SLCA and LCC and allows for a screening assessment of all three dimensions of sustainability. Meanwhile, it lowers the entry barrier to implementing basics of LCSA in industry and communicating with non-expert practitioners.

Tier 2 represents a "best practice" of LCSA considering additional categories (e.g. the common used ones currently considered in the ILCD Handbook (JRC 2010b) of LCA and categories for SLCA and LCC, which have been ranked as important. Hence, additional impact categories for LCA, for example ozone depletion, eutrophication, photochemical oxidant formation, acidification, have been chosen. For LCC, consumer costs (e.g. purchase price, maintenance costs and energy costs) are included. For SLCA, health (including workers, consumers and local communities) and working conditions are taken into account. Thus, Tier 2 provides a broader range of environmental and economic aspects, and includes social topics beyond the stakeholder group workers.

The most advanced step, Tier 3, represents a comprehensive level of LCSA considering a broad set of categories (e.g. for potential new LCA impact categories like water footprint methods and land use). For LCC, production and consumer

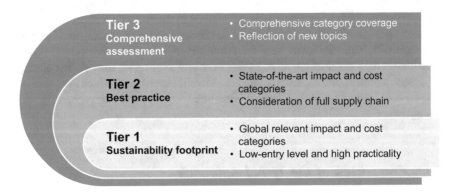

Fig. 2 Structure of the Tiered Approach—3 tiers reflecting different levels of comprehensiveness of LCSA (Neugebauer et al. 2015)

costs related to further operation, accidents, and environmental damage (if not considered within LCA and SLCA) are considered. For SLCA, the topics education, human rights, and cultural heritage are addressed.

The Tiered Approach supports a holistic sustainability assessment, as all three dimensions of sustainability are considered. In addition, it ensures practicality through its impact and cost categories hierarchy, reflecting different levels of comprehensiveness and different phases of LCSA's development.

3.2 Implementation in Manufacturing

The practicality of the Tiered Approach has been proven by first case studies on manufacturing technologies and products, e.g. modular machine tool frames and wireless micro systems (Peukert et al. 2015; Benecke et al. 2015), turning technologies as well as bicycles and pedal electric cycles (Neugebauer et al. 2013; Buchert et al. 2015). The case studies mainly focused on the Tier 1, i.e. the categories climate change, production costs and fair wages. They revealed environmental hotspots, described first selected social topics (e.g. wages) and first economic issues (e.g. production costs), and identified improvement potential for these technologies and products.

Meanwhile, by carrying out these case studies, knowledge and experience with regard to practical implementation were gained from the identification of hotspots and the interpretation of life cycle impacts of the three sustainability dimensions. Specific social aspects for example, fair wages and health, were mapped and thus compared for different countries involved in the production of smart modular machine tool frames, e.g. Germany, Brazil, and China (Peukert et al. 2015). Based on the results, recommendations could be given for advantageous material usage, supplier management and further technology improvements.

Moreover, trade-offs between the three sustainability dimensions were identified, e.g. a technology which performs better from an environmental perspective, could however lead to higher social risks. For instance, the switch from wet machined turning processes to inner-cooled ones showed potential environmental benefits (e.g. recycling of titanium chips), but at the same time increased the social risk due to the African workers involved in the inlay production being potentially paid below the poverty line.

3.3 Research Needs and Outlook

The Tiered Approach is a first step with regard to fostering LCSA in practice. However, challenges remain as comprehensive category sets as well as well-defined impact pathways for all three tiers are missing in the case of both SLCA and LCC. Moreover, at the interpretation phase, challenges occur due to the potential trade-off of the results between and within the three sustainability dimensions (Zamagni et al. 2013; Arcese et al. 2013). In the case studies described above, those trade-offs were displayed transparently for each dimension in the Tiered Approach without giving weights.

The next steps will focus on updating the selected categories and the hierarchy of the Tiered Approach, and on developing impact pathways for social and economic aspects suitable for LCSA with regard to production technologies.

4   Selected Ongoing LCSA Work

Currently, many studies have been carried out in pursuit of enhancing implementation, scientific robustness, and comprehensiveness of the three methods with LCSA. In this section, some ongoing work has been selected to show the recent research progress and direction of LCSA development particularly with regard to SLCA and LCC.

4.1   Proposals of Impact Pathways for SLCA and LCC

As described in the previous sections, SLCA and LCC face numerous challenges, particularly with regard to the impact assessment stage, which hinder the implementation and methodological robustness of LCSA. This includes missing concrete impact category definitions of SLCA and LCC, missing detailed impact pathways, as well as insufficient description of the relationship between impact categories and AoPs (Bocoum et al. 2015; Chhipi-Shrestha et al. 2015; Andreas Jørgensen et al. 2008; Neugebauer et al. 2014). First steps to address these gaps were done by establishing first impact pathways for the social dimension, describing the relation between indicators and impact categories with a focus on fair wage and level of education (Neugebauer et al. 2014), and by proposing AoPs for the social and the economic dimension, such as social justice and economic stability (Neugebauer et al. 2016; Neugebauer et al. 2014). The development of the impact pathways is introduced in the following section.

4.1.1   Proposal of Social Impact Pathways: Fair Wage and Level of Education

In order to enhance SLCA and thus LCSA, impact categories need to be clearly defined. Furthermore, impact pathways linking indicators to impact categories and AoPs need to be developed.

To that end, Neugebauer et al. (2014) defined two impact categories at a midpoint level and developed social impact pathways for them. The two topics are recognised as essential aspects for Sustainable Development Goals (United Nations 2016) for mitigating poverty and enabling the achievement of higher prosperity levels. In manufacturing, fair wage is treated as an essential aspect of worker's overall living situation and well-being. Education reflects country-specific equality aspects, and measures worker's qualifications for specific sectors and countries. With the development of the two midpoint categories, three related endpoint categories (environmental stability, damage to human health, and economic welfare) and two AoPs (social well-being and social justice) were proposed to complete the impact pathways. Interrelations along the defined pathways have been introduced, e.g. the inventory indicator lowest/highest gross income affects the AoPs social justice and social well-being through the midpoint impact category fair wage and the endpoint impact categories economic welfare and damage to human health. Similar to the impact pathway for fair wage, the relation of the inventory indicators, such as access barriers to schools, to the midpoint impact level of education, was investigated.

The proposal of potential impact pathways of fair wage and level of education, serves to facilitate a more consistent and transparent assessment of social impact. However, the characterisation factors stay at a qualitative level. The next step for refining the impact pathways focuses on the identification of quantitative characterisation factors instead of purely on qualitative descriptions. Further aspects like the interpretation of social impacts have been investigated in tandem.

4.1.2 Introduction of the New Economic Life Cycle Assessment Framework

As pointed out in Sect. 2.2, LCC so far includes pure cost assessment without considering clearly defined AoPs, impact categories and corresponding causalities described in impact pathways. For this reason, some authors discuss whether LCC can actually adequately measure the economic sustainability dimension within the LCSA framework (Jørgensen et al. 2010; Heijungs et al. 2013).

Taking into account this discussion, Neugebauer et al. (2016) proposed the new Economic LCA (EcLCA) framework, which broadens the scope of the current LCC by including the impact assessment stage. As a result, two AoPs (economic stability and wealth generation), two endpoint impact categories (economic prosperity and economic resilience), and five midpoint impact categories (profitability, productivity, consumer satisfaction, business diversity, and long-term investment) are suggested and defined. The proposed midpoint impact categories can be directly linked to manufacturing. For example, profitability considers costs regarding actual economic benefits for the firms via added values instead of purely summing up costs. Furthermore, productivity is associated with human capital aspects through the whole value chains, and consumer satisfaction influences the markets and product management expenses, etc.

The suggested EcLCA framework better meets the requirements of ISO 14040 (ISO 2006a) and 14044 (ISO 2006b) adopted within the LCSA framework and describes economic aspects targeting sustainability. The next steps would be to establish measurable linkages (i.e. quantitative relation) between inventory and impact levels as well as AoPs. Moreover, trials for testing application of the new framework will constitute part of future work.

4.2 Sustainability Safeguard Star

LCSA considers the three dimensions of sustainability by combining the methods LCA, LCC and SLCA. However, there is a risk that social, environmental and economic aspects are only interpreted individually, without considering potential interlinkages between the sustainability dimensions. For instance, climate change impacts influence AoPs in both SLCA and LCA, i.e. social well-being (e.g. by affecting human health) and ecosystem quality. To address this challenge, the Sustainability Safeguard Star was designed to structure existing AoPs used in LCA into a new scheme by addressing the inter-linkages in between the three sustainability dimensions and by including additional topics of sustainability, such as social justice (Schmidtz 2006; Neugebauer et al. 2014) and economic stability (Neugebauer et al. 2016). The proposed framework is introduced in the following section.

4.2.1 Conceptual Framework of Sustainability Safeguard Star

The Sustainability Safeguard Star goes beyond the three broadly accepted AoPs from the classical (environmental) LCA human health, resource availability, and ecosystem quality (JRC 2010a), with the goal of defining common AoPs for the LCSA framework. This means that the Sustainability Safeguard Star additionally considers three complementary AoPs (i.e. safeguard subjects), which then reflect the social and economic dimension of sustainability: man-made environment, social justice, and economic stability. The six AoPs proposed for LCSA are displayed in Fig. 3.

The AoP man-made environment, which was already proposed by de Haes et al. (1999), stands for cultural value and addresses technical infrastructure, such as energy and communication networks, and the drinking water supply, indicating the living contexts of society. The AoP is, for example, concerned with the damage resulting from acidifying substances to buildings. The other AoP, social justice, takes equal opportunities and justice as core principles, like security of freedom based on a social contract (individual vs. societal). It is of high relevance to address social justice (Nussbaum 2004) issues in order to eliminate inequality, foster human rights and intergenerational equity defined as fundamental to sustainable development pursuits as defined by the Brundtland report (United Nations 1987). Last but not least, another AoP, economic stability, aims at avoiding economic crisis and promoting economic growth and employment (European Commission 2014). It is also connected to industrial diversity and multilateral trade concerns for addressing economic vulnerability (Neugebauer et al. 2016). The AoPs defined combine different aspects to consider interlinkages between the sustainability dimensions. The AoP economic stability, for example, addresses unemployment and economic prosperity, which are associated with both social and economic perspectives.

Fig. 3 Sustainability Safeguard Star: conceptual framework and relation to LCSA

Moreover, Fig. 3 shows the general conceptual framework for the potential links between micro- and macroeconomic level. The proposed AoPs reflect sustainability goals at a macroeconomic level (e.g. from sustainable development goals or strategies defined by United Nations (2016) and European Commission (2010)). These goals, for example, reducing inequality, can be assessed by defined criteria (e.g. equal access to all levels of education). With the inclusion of the proposed AoPs and their impact pathways addressing the defined criteria, LCSA can deliver the results at the microeconomic level.

4.2.2 Research Needs and Outlook

The Sustainability Safeguard Star abolishes the presumed separation of AoPs defined in three underlying life cycle methods of LCSA and in their place, suggests six common AoPs which address the inter-linkages in between the three sustainability dimensions.

Further research should focus on establishing impact pathways between defined impact categories and the proposed AoPs (see also Sects. 4.1.1 and 4.1.2) and tested in case studies. With regard to sustainable manufacturing, the newly defined AoPs of economic stability and man-made environment, can be of relevance for the purpose of reflecting the business situation of firms with the background of different production locations.

5 Conclusion

The Life Cycle Sustainability Assessment (LCSA) framework is applied to assess the sustainability performances of manufacturing products and processes. Application of LCSA can lead to the identification of product and process hotspots, and support decision-making in production development. In favour of implementation of LCSA in practice, the Tiered Approach was proposed to provide an impact and cost category hierarchy, particularly for offering guidance to practitioners in industry. This approach has already been applied in first case studies on manufacturing technologies and products, e.g. turning technologies and pedal electric cycles, and has proven its validity. Ongoing work such as the development of impact pathways for SLCA, the suggested Economic LCA, and the Sustainability Safeguard Star, serve to enhance the robustness and applicability of the LCSA. To continue enhancing currently proposed methods, future work need to focus on developing the impact pathways of economic and social aspects in the context of LCSA, and further providing quantitative measures of the pathways.

References

Andreas Jørgensen, 2013. Social LCA—a way ahead? *International Journal of Life Cycle Assessment* 18(2): 296–299. doi:10.1007/s11367-012-0517-5.

Andreas Jørgensen, Agathe Le Bocq, Liudmila Nazarkina, and Michael Hauschild. 2008. Methodologies for social life cycle assessment. *International Journal of Life Cycle Assessment* 2: 96–103. http://www.me.mtu.edu/~jwsuther/colloquium/Nesbitt_MethodolgiesforSLCA.pdf.

Andreas Jørgensen, Lufanna C.H. Lai, and Michael Z. Hauschild. 2010. Assessing the validity of impact pathways for child labour and well-being in social life cycle assessment. *International Journal of Life Cycle Assessment* 15: 5–16. doi:10.1007/s11367-009-0131-3.

Andreas Jørgensen, Michael Z. Hauschild, Michael S. Jørgensen, and Arne Wangel. 2009. Relevance and feasibility of social life cycle assessment from a company perspective. *International Journal of Life Cycle Assessment* 14(3): 201–214. doi:10.1007/s11367-009-0073-9.

Arcese, Gabriella, Roberto Merli, and Maria Claudia Lucchetti. 2013. Life cycle approach : A critical review in the tourism sector. In *The 3rd World Sustainability Forum*, 1–10.

Bach, Vanessa, and Matthias Finkbeiner. 2016. Approach to qualify decision support maturity of new versus established impact assessment methods—demonstrated for the categories acidification and eutrophication. *International Journal of Life Cycle Assessment* (accepted). *The International Journal of Life Cycle Assessment*. doi:10.1007/s11367-016-1164-z.

Bare, Jane C., Patrick Hofstetter, David W. Pennington, and Helias a. Udo Haes. 2000. Midpoints versus endpoints: The sacrifices and benefits. *International Journal of Life Cycle Assessment* 5(6): 319–326. doi:10.1007/BF02978665.

Benecke, Stephan, Sabrina Neugebauer, Bernd Peukert, and et al. 2015. Sustainability assessment of wireless micro systems in smart manufacturing environments. EcoDesign 2015. Tokyo, Japan. (Siehe Anhang). In *EcoDesign 2015, Tokyo, Japan*.

Benoît, Catherine, Marzia Traverso, Sonia Valdivia, Gina Vickery-Niederman, Juliane Franze, Lina Azuero, Andreas Ciroth, Bernard Mazijn, and Deana Aulisio. 2013. The methodological sheets for sub-categories in social life cycle assessment (S-LCA).

Bocoum, Ibrahima, Catherine Macombe, and Jean-Pierre Revéret. 2015. Anticipating impacts on health based on changes in income inequality caused by life cycles. *International Journal of Life Cycle Assessment* 20(3): 405–417. doi:10.1007/s11367-014-0835-x.

Buchert, Tom, Jón Garðar Steingrímsson, Sabrina Neugebauer, The Duy Nguyend, Mila Galeitzkeb, Nicole Oertwig, Johannes Seidel, et al. 2015. Design and manufacturing of a sustainable pedelec. In *The 22nd CIRP Conference on Life Cycle Engineering, Procedia CIRP* 29: 579–84.

Chhipi-Shrestha, Gyan Kumar, and Rehan Sadiq Kasun Hewage. 2015. 'Socializing' sustainability: A critical review on current development status of social life cycle impact assessment method. *Clean Technologies and Environmental Policy* 17(3): 579–596. doi:10.1007/s10098-014-0841-5.

Dreyer, Louise Camilla, Michael Z. Hauschild, and Jens Schierbeck. 2006. A framework for social life cycle impact assessment. *International Journal of Life Cycle Assessment* 11(2): 88–97. doi:10.1065/lca2005.08.223.

Elkington, John. 1998. *Cannibals with forks: The triple bottom line of 21st century business*. Gabriola Island: New Society Publishers.

European Commission. 2010. EUROPE 2020—a strategy for smart, sustainable and inclusive growth. Brussels, Belgium. doi:10.1016/j.resconrec.2010.03.010.

European Commission. 2014. Economic stability and growth. *European Commission*. http://ec.europa.eu/economy_finance/euro/why/stability_growth/index_en.htm.

European Commission. 2015. European platform on life cycle assessment (LCA).

Finkbeiner, Matthias, Erwin M. Schau, Annekatrin Lehmann, and Marzia Traverso. 2010. towards life cycle sustainability assessment. *Sustainability* 2: 3309–3322. www.mdpi.com/2071-1050/2/10/3309/pdf.

Finkbeiner, Matthias, Robert Ackermann, Vanessa Bach, Markus Berger, Gerhard Brankatschk, Ya-Ju Chang, Marina Grinberg, et al. 2014. Challenges in life cycle assessment: An overview of current gaps and research needs. In *Background and future prospects in life cycle assessment*, ed. Walter Klöpffer, 207–58. Springer. doi:10.1007/978-94-017-8697-3_7.

Glucha, Pernilla, and Henrikke Baumann. 2004. The life cycle costing (LCC) approach: A conceptual discussion of its usefulness for environmental decision-making. *Building and Environment* 39(5): 571–580. doi:10.1016/j.buildenv.2003.10.008.

Heijungs, Reinout, Ettore Settanni, and Jeroen Guinée. 2013. Toward a computational structure for life cycle sustainability analysis: Unifying LCA and LCC. *International Journal of Life Cycle Assessment* 18(9): 1722–1733. doi:10.1007/s11367-012-0461-4.

Hunkeler, David, Kerstin Lichtenvort, and Gerald Rebitzer. 2008. *Environmental life cycle costing*. Boca Raton, USA: CRC Press.

ISO. 2006a. ISO 14040:2006 environmental management—life cycle assessment—principles and framework. Geneva (Switzerland): ISO.

ISO. 2006b. ISO 14044:2006 environmental management—life cycle assessment—requirements and guidelines. Geneva (Switzerland): ISO.

ISO. 2008. ISO 15686-5:2008. Buildings and constructed assets—service-life planning—Life Cycle costing. Geneva (Switzerland): ISO.

Jørgensen, Andreas, Ivan T. Hermann, and Jørgen Birk Mortensen. 2010. Is LCC relevant in a sustainability assessment? *International Journal of Life Cycle Assessment* 15(6): 531–532. doi:10.1007/s11367-010-0185-2.

JRC. 2010a. ILCD handbook—framework and requirements for LCIA models and indicators. *Luxembourg*. doi:10.2788/38719.

JRC. 2010b. ILCD handbook—general guide for life cycle assessment—detailed guidance. *Luxembourg*. doi:10.2788/38479.

JRC. 2011. ILCD handbook—recommendations for life cycle impact assessment in the European context. First. *Ispra: European Commission*. doi:10.278/33030.

Klöpffer, Walter. 2008. Life cycle sustainability assessment of products. *International Journal of Life Cycle Assessment* 13(2): 89–95. doi:10.1065/lca2008.02.376.

Lehmann, Annekatrin. 2013. Lebenszyklusbasierte Nachhaltigkeitsanalyse von Technologien: Am Beispiel Eines Projekts Zum Integrierten Wasserressourcenmanagement. Ph.D. Dissertation, Technische Universität Berlin.

Lehmann, Annekatrin, Eva Zschieschang, Marzia Traverso, Matthias Finkbeiner, and Liselotte Schebek. 2013. Social aspects for sustainability assessment of technologies—challenges for social life cycle assessment (SLCA). *International Journal of Life Cycle Assessment* 18(8): 1581–1592. doi:10.1007/s11367-013-0594-0.

Martínez-Blanco, Julia, Annekatrin Lehmann, Pere Muñoz, Assumpció Antón, Marzia Traverso, Joan Rieradevall, and Matthias Finkbeiner. 2014. Application challenges for the social LCA of fertilizers within life cycle sustainability assessment. *Journal of Cleaner Production, January.*. doi:10.1016/j.jclepro.2014.01.044.

May, J.R., and D.J. Brennan. 2006. Sustainability assessment of australian electricity generation. *Process Safety and Environmental Protection* 84(2): 131–142. doi:10.1205/psep.04265.

Neugebauer, Sabrina, Ya-Ju Chang, Markus Maliszewski, Kai Lindow, Rainer Stark, and Matthias Finkbeiner. 2013. Life cycle sustainability assessment & sustainable product development: A case study on pedal electric cycles (Pedelec). In *Proceedings of 11th global conference on sustainable manufacturing, September 23-25, Berlin, Germany*. ISBN: 978-3-7983-2608-8.

Neugebauer, Sabrina, Silvia Forin, and Matthias Finkbeiner. 2016. From life cycle costing to economic life cycle assessment—introducing an economic impact pathway. *Sustainability* 8(5): 1–23. doi:10.3390/su8050428.

Neugebauer, Sabrina, Julia Martinez-blanco, René Scheumann, and Matthias Finkbeiner. 2015. Enhancing the practical implementation of life cycle sustainability assessment—proposal of a tiered approach. *Journal of Cleaner Production* 102: 165–176. doi:10.1016/j.jclepro.2015.04.053.

Neugebauer, Sabrina, Marzia Traverso, René Scheumann, Ya-ju Chang, Kirana Wolf, and Matthias Finkbeiner. 2014. Impact pathways to address social well-being and social justice in SLCA—fair wage and level of education. *Sustainability* 6(8): 4839–4857. doi:10.3390/su6084839.

Nussbaum, Martha C. 2004. Beyond the social contract: Capabilities and global justice. *Oxford Development Studies* 32(1): 3–18. doi:10.1080/1360081042000184093.

Öko-Institut. 1987. Product portfolio analysis—needs, products and consequences (Produktlinienanalyse - Bedürfnisse, Produkte Und Ihre Folgen). Kölner Volksblatt Verlag.

Peukert, Bernd, Stephan Benecke, Janire Clavell, Sabrina Neugebauer, Nils F. Nissen, Eckart Uhlmann, Klaus-Dieter Lang, and Matthias Finkbeiner. 2015. Addressing sustainability and flexibility in manufacturing via smart modular machine tool frames to support sustainable value creation. In *The 22nd CIRP conference on life cycle engineering, Procedia CIRP* 29: 514–19.

Rainer Grießhammer, Matthias Buchert, Carl-Otto Gensch, Christian Hochfeld, Andreas Manhart, and Ina Rüdenauer. 2007. *PROSA—product sustainability assessment*. Öko-Institut e.V.

Remmen, Arne, Allan Astrup, and Jeppe Frydendal. 2007. *Life cycle management—a business guide to sustainability*. UNEP/SETAC Life Cycle Initiative.

Wood, Richard, and Edgar G. Hertwich. 2012. Economic modelling and indicators in life cycle sustainability assessment. *International Journal of Life Cycle Assessment.*

Schmidtz, David. 2006. *The elements of justice.* Cambridge, UK: Cambridge University Press.

Swarr, T., D. Hunkeler, W. Kloepffer, H.L. Pesonen, A. Ciroth, C. Brent, and R. Pagan. 2011. *Environmental life cycle costing: A code of practice.*

de Haes, Udo, A. Helias, Olivier Jolliet, Göran Finnveden, Michael Hauschild, Wolfram Krewitt, and Ruedi Miiller-Wenk. 1999. Best available practice regarding impact categories and category indicators in life cycle impact assessment. *The International Journal of Life Cycle Assessment* 4(3): 167–174. doi:10.1007/BF02979453.

UNEP. 2012. *Greening the economy—through life cycle thinking.* Paris: UNEP.

UNEP/SETAC. 2009. *Guidelines for social life cycle assessment of products.* Belgium: In de Weer.

UNEP/SETAC Life Cycle Initiative. 2011. *Towards a life cycle sustainability assessment—making informed choices on products.* Paris: United Nations Environment Programme.

United Nations. 1987. Report of the World Commission on Environment and Development: Our common future. United Nations. http://www.un-documents.net/wced-ocf.htm.

United Nations. 2016. Sustainable development goals. https://sustainabledevelopment.un.org/sdgs.

Valdivia, Sonia, Cassia M.L. Ugaya, Jutta Hildenbrand, Marzia Traverso, Bernard Mazijn, and Guido Sonnemann. 2012. A UNEP/SETAC approach towards a life cycle sustainability assessment—our contribution to Rio + 20. *International Journal of Life Cycle Assessment* 18 (9): 1673–1685. doi:10.1007/s11367-012-0529-1.

Zamagni, Alessandra, Hanna-Leena Pesonen, and Thomas Swarr. 2013. From LCA to life cycle sustainability assessment: Concept, practice and future directions. *International Journal of Life Cycle Assessment* 18(9): 1637–1641. doi:10.1007/s11367-013-0648-3.

Optimisation Methods in Sustainable Manufacturing

Sebastian Schenker, Ingmar Vierhaus, Ralf Borndörfer,
Armin Fügenschuh and Martin Skutella

1 Introduction

Sustainable manufacturing is driven by the insight that the focus on the economic dimension in current businesses and lifestyles has to be broadened to cover all three pillars of sustainability: economic development, social development, and environmental protection. In this chapter, we present two state-of-the-art approaches of mathematical optimisation and how they can be used to solve problems in sustainable manufacturing.

The multi-criteria perspective considers areas of sustainability as independent functions that are to be optimised however with divergent objectives simultaneously. Accordingly, computed outcomes that cannot be improved upon (on at least one objective without getting worse at another) are considered to be superior to outcomes that can be improved upon. A decision maker will only be interested in the first set of outcomes in order to be able to form an educated opinion with respect to his/her sustainability goal.

The system dynamics perspective on the other hand focuses on the time-dependent (or dynamic) aspects of systems that are influenced by sustainable manufacturing practices. If, for instance, a production technology was identified

S. Schenker (✉) · I. Vierhaus · R. Borndörfer
Zuse Institute Berlin, Takustr. 7, 14195 Berlin, Germany
e-mail: schenker@zib.de

A. Fügenschuh
Helmut-Schmidt-University, Holstenhofweg 85, 22043 Hamburg, Germany

M. Skutella
TU Berlin, Str. des 17. Juni 136, 10623 Berlin, Germany

R. Stark et al. (eds.), *Sustainable Manufacturing*, Sustainable Production,
Life Cycle Engineering and Management, DOI 10.1007/978-3-319-48514-0_15

that cannot be improved in either of the sustainability dimensions, the question then arises as to how this technology can be used in an optimal way using only limited resources. How can the impact on society and economy be steered in the direction of allowing the technology to be as beneficial as possible?

2   Multi-criteria Optimisation

Mathematical optimisation and mathematical programming is concerned with finding good solutions from a set of available alternatives. The abstract nature of mathematical optimisation allows the user to model a wide range of different problems and different objectives using the same theoretical insights and practical tools. Problems in sustainability and sustainable manufacturing have in common that there is not only one objective to be considered but several conflicting ones. This is mathematically reflected by considering several objective functions simultaneously. The set of available alternatives and the structure of the considered objective functions can generally be modelled in different ways. The focus in the following section is put on the well-studied and fruitful field of linear optimisation involving linear objective functions and linear constraints allowing the user to model as well as to efficiently solve a wide range of quantitative problems.

2.1   Multi-criteria Problem Formulation

In a *general* multi-criteria linear optimisation problem, one is given a set of k cost vectors $c_1, \ldots, c_k \in \mathbb{R}^n$ and seeks to minimize all linear cost functions $c_i \cdot x = \sum_{j=1}^{n} c_{ij} x_j$, for $i = 1, \ldots, k$, simultaneously over all n-dimensional vectors $x = (x_1, \ldots, x_n)$ subject to a set of linear inequality and integer constraints. In particular, let M be some finite index set and suppose that for every $i \in M$, we are given an n-dimensional vector a_i and a scalar b_i. Let N_1, N_2 and N_3 be subsets of $\{1, \ldots, n\}$ that indicate which variables x_j are constrained to be non-negative, binary or integer, respectively. We then consider the problem

$$
\begin{aligned}
\min(c_1 \cdot x, &\ldots, c_k \cdot x) \\
\text{s.t. } a_i \cdot x \le b_i, &\qquad i \in M, \\
x_j \ge 0, &\qquad j \in N_1, \\
x_j \in \{0, 1\}, &\qquad j \in N_2, \\
x_j \in \mathbb{Z}, &\qquad j \in N_3.
\end{aligned} \tag{1}
$$

Fig. 1 Feasible space of a bi-criteria integer maximization problem and corresponding set in objective space with non-dominated points (*red*)

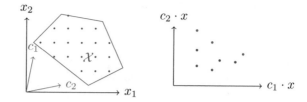

The variables $x_1, \ldots, x_n$ are called *decision variables* and a vector x satisfying all of the constraints is called a *feasible solution*. The set of all feasible solutions is called *feasible set* and will be denoted by $\mathcal{X}$. The image $y = (c_1 \cdot x, \ldots, c_k \cdot x)$ of a feasible solution x is called a *feasible point* and the set of all feasible points is called objective set and will be denoted by $\mathcal{Y}$. If N_1 coincides with $\{1, \ldots, n\}$ (implying $N_2 = N_3 = \emptyset$), then (1) is considered a *linear* programming problem. If $N_2 = \{1, \ldots, n\}$ or $N_3 = \{1, \ldots, n\}$, then we refer to (1) as a *binary* or *integer* programming problem, respectively. In case of $\emptyset \subsetneq N_1 \subsetneq \{1, \ldots, n\}$, (1) is considered a *mixed-integer* programming problem. The earliest investigations of multicriteria mathematical optimisation go back to the 1950s when the simplex method coined by Dantzig opened up a wide range of applications and prepared the ground for the huge success of linear programming (Dantzig 1963). If $k = 1$, then we refer to (1) as a single-objective problem and the notion of optimality is unambiguous. For a multi-criteria optimisation problem (with number of objectives $k \geq 2$) we cannot expect to find a solution that optimizes all objectives simultaneously leading to several possible notions of optimality in the multi-criteria case (Ehrgott 2005). A widely accepted (and in the following considered) one is the notion of *efficiency*. A solution $x^* \in \mathcal{X}$ is considered *efficient* if there is no other solution $x \in \mathcal{X}$ that achieves objective values at least as good with a strictly better value in at least one objective, i.e., there is no $x \in \mathcal{X}$ with $c_i \cdot x \leq c_i \cdot x^*$ for $i = 1, \ldots, n$ and $c_i \cdot x < c_i \cdot x^*$ for at least one $i \in \{1, \ldots, n\}$. The image of an efficient solution is called *non-dominated*. The challenge for a multi-criteria optimisation problem is then to compute all different non-dominated points (Figs. 1 and 2).

Fig. 2 Feasible space of a bi-criteria linear maximization problem and corresponding set in objective space with non-dominated points (*red*)

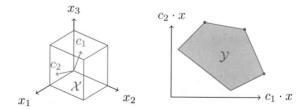

2.2 Manufacturing and Scheduling

Production problems, scheduling problems and similar decision problems are a fruitful domain for (mixed) integer programming. Binary variables might represent on-off decisions and linear or integer variables, respectively, might represent production quantities. In the following we will shortly present how multi-criteria integer programming could be used to model a scheduling problem that accounts for production costs, electricity consumption and worker satisfaction. Lets $[M] = \{1,\ldots,M\}$ be a finite set representing a set of different machines and let $[J] = \{1,\ldots,J\}$ be a finite set representing a set of jobs. We will consider a time horizon for the entire production process and let s and e be the start and end time of it. Introduce variables $x_{jmt} \in \{0,1\}$ where $j \in [J]$, $m \in [M]$ and $t \in \{s,\ldots,e\}$. We set $x_{jmt} = 1$ if and only if starting time of job j on machine m is set to t. In order to model the constraint that every job needs to run on every machine before end time e, let $dur(m)$ the duration on machine m, i.e., the time that a job spends on machine m. Then,

$$\sum_{t=s}^{e-dur(m)} x_{jmt} = 1 \; \forall j \in [J] \wedge \forall m \in [M] \tag{2}$$

models the above fulfilment constraint. Furthermore, the constraint that job j is only allowed to run on machine $m+1$ if it is finished on machine m can be modelled via

$$\sum_{t=s}^{e} t \cdot x_{jmt} + dur(m) \leq \sum_{t=s}^{e} t \cdot x_{jm+1t} \forall j \in J \wedge \forall m \in [M-1] \tag{3}$$

Furthermore, it is very reasonable to assume that a new job can only be started on machine m if the previous job on machine m was finished. This constraint could be modelled via

$$\sum_{t=s}^{e} t \cdot x_{jmt} + dur(m) \leq \sum_{t=s}^{e} t \cdot x_{j+1mt} \forall j \in [J-1] \wedge \forall m \in [M] \tag{4}$$

2.3 Solving Multi-criteria Optimisation Problems

For the single-objective case there are several commercial solvers and software packages (CPLEX 2016; Xpress 2016; Gurobi 2016) and non-commercial ones (Achterberg 2009). One could have expected that the exponential growth in computing power and the even larger algorithmic speed-ups in mixed integer programming during the last decade (Bixby 2002) would automatically lead to multi-criteria extensions. But the situation is contrary: none of the available

Fig. 3 Front of
non-dominated points for a
bi-criteria bicycle
manufacturing problem

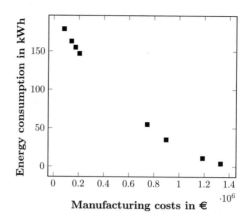

commercial solvers supports multi-criteria problems and there are only a few, recently developed non-commercial solvers available: BENSOLVE (Löhne and Weiing 2014) and inner (Csirmaz 2016) handle multi-criteria linear programming problems, SYMPHONY (Ladanyi et al. 2016) supports bi-criteria mixed integer problems and PolySCIP (Schenker et al. 2016) supports multi-criteria linear and integer problems.

PolySCIP reads problems of the above form (1) via its MOP file format which is based on the widely used MPS file format (MPS-Format 2016) and allows the user to model constraints like (2), (3), (4) easily via an algebraic modelling language (Koch 2004). It can handle an arbitrary number of objectives and thousands of variables and constraints (Fig. 3).

3 System Dynamics Optimisation

In this book, many technologies and approaches developed in the context of sustainable manufacturing are discussed. In this section, we will consider the global environment in which these technologies must be disseminated and implemented, in order to realise their positive potential.

The economy, the environment, and the society constitute complex entities and can be seen as finely balanced networks of mutual dependencies. Almost all components influence each other that have either supporting or weakening effects. Such dynamical systems can demonstrate counterintuitive behaviour. However, in order to bring about a change from the conventional production paradigm in the direction of a paradigm of sustainability, it is essential to appreciate the complex interdependencies of the systems involved.

We observe that the transition, i.e., the setup of many value creation modules and networks, constitutes a dynamic process over time that will span several years or decades. During this period, an array of interactions between the stakeholders need to be taken into account. Moreover, the transition does not take place by itself. It will only happen by means of deliberate influence on the system. A bundle of individual measures are necessary in this process.

To this end, the system dynamics (SD) approach provides the appropriate framework. It is an approach for the modelling and simulation of dynamical systems with a long history rooted in the understanding and teaching of dynamical systems in general, as well as in the field of sustainability.

After introducing system dynamics as a tool for simulation, we will formulate optimal control problems based on system dynamics models.

3.1 System Dynamics

In this section, we will introduce system dynamics as a modelling methodology as well as the most important modelling rules and characteristics of system dynamics models.

System dynamics was introduced by Jay Forrester in the 1950s as a method of describing and simulating time-dependent effects of complex influence networks with feedback loops (Forrester 1961). Such networks are characterized by non-linear, often surprising behaviour. In fact, a forecast of their future development, and thus their control, represents a difficult mathematical problem.

One of the strengths of the system dynamics approach lies in its visual representation of complex systems. This visual approach is essential in the system dynamics modelling process, and simplifies access for beginners and users who lack experience with systems of differential equations.

The main objects of system dynamics models are *stocks* and *flows*. The stocks contain the state information of the system. By convention, each stock has two flows, one flowing into the stock, and one flowing out of the stock. Figures 4 and 5 show visual representations of a stock and a flow respectively. As a third component, *auxiliary variables* are often introduced to structure a diagram. Lastly, the existence of functional dependencies between stocks, flows and variables is indicated by arrows. Figure 6 shows an example.

Using this visual representation, a systematic modelling process could be structured as follows:

- Definition of the modelling goal,
- Definition of the system limits,
- Definition of the system components,
- Definition of the direct relations between system components and the type of causal links (positive or negative),
- Design of an influence diagram to summarize components and their relations,

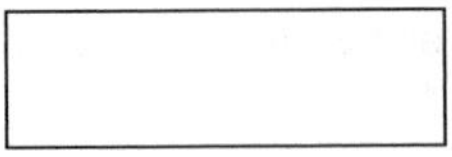

Fig. 4 Visual representation of a stock

Fig. 5 Visual representation of a flow. The origin of the flow is outside of the limits of the system, as indicated by the *cloud symbol*. The *arrow* is decorated by an hourglass to indicate time dependency

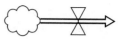

Fig. 6 Visual representation of a small model with one stock, one flow and one variable. The value of the flow depends on the value of the variable, which in turn depends on the value of the stock

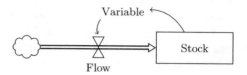

- Creation of a system dynamics diagram with stocks for each of the system components as well as flows for each stock,
- Assignment of units and valid ranges to the values of stocks and flows,
- Definition of the functional relations between stocks and flows,
- Introduction of variables to simplify the relations if possible,
- Completion of the system dynamics diagram by adding variables and arrows for relations,

The result of this process is a complete system dynamics model. In the next section, we will discuss numerical methods for simulating a system dynamics model as it develops over time.

Although it is possible to find general solutions analytically for some models, this is generally neither possible nor required. A range of numerical simulation techniques exist that provide quickly accurate simulations. One class of such simulation techniques are the Runge-Kutta schemes (Runge 1895; Kutta 1901) which we will use in this chapter.

3.2 Optimal Control of System Dynamics Models

As we discussed in the previous sections, in its basic form, SD aims at describing and simulating influence networks. This is an important step in pursuit of understanding the mutual dependencies. In addition to obtaining a mere understanding however, what we would like to do is to intervene in the network, bring it to a desired stable state, or get as close as possible to that state.

In system dynamics, the points of the system which can be influenced by a conscious decision of an actor are modeled using the concept of *policies*.

Policies constitute a basic and important concept of system dynamics modelling. A policy is a function in some variables that describes the rates of flow in a system and hence the dynamic behaviour of the model (Richardson and Pugh 1981). Thus, a policy is a decision rule which specifies how a decision-maker processes available information from model variables (Sterman 2000). Questions regularly arise concerning whether a given policy can be improved, or even what a "good" policy "actually constitutes or entails. In this context, the need for efficient computational methods for policy analysis as well as policy improvement and design has been recognized in system dynamics, see, e.g., Yücel and Barlas (2011), Keloharju and Wolstenholme (1988), and is an active field of research.

When developing a simulation model, the modelling step of "policy formulation and evaluation" also compares the performance of two or more candidate policies (Sterman 2000). When two simulations with different policies lead to different system behaviors, one has to evaluate which of the two simulations is more suitable or "better" for a given model purpose. To answer this question, one needs to define an objective function so that the higher the value of the objective function for a given simulation, the more favorable or "better" the policy (Dangereld and Roberts 1996). Once an objective function is defined, several approaches to computer-aided policy improvement are at one's disposal.

Direct parameter policy design starts with the definition of an analytic, parametrized, and usually nonlinear policy function (Keloharju and Wolstenholme 1989). The parameters of this function are set to starting values, and for each parameter, a range of valid values is defined. These parameters constitute then the free variables of the optimisation problem, i.e., the variables which can be varied freely in pursuit of an optimal solution. Consequently, the goal of the policy improvement is to find a set of parameter values within the given range that improves the value of the objective function. The solution space in this case is reduced by the *a priori* definition of the shape of the policy function. The solution found by the optimisation algorithm depends strongly on this definition and therefore on the expectations of the modeler. If a software package offers parameter optimisation capabilities, it is usually possible to attempt producing the solution of such direct parameter policy design problems.

Table function policy design is one possible way to generalizing direct parameter policy design, by defining a parametrized table function instead of an analytic function (Keloharju and Wolstenholme 1989). In this case, the modeler has to define the number of data points of the table function and two intervals that define the range of valid values of the data points on the x- and y-axis. This approach removes the modeler's expectations of the shape of the policy from the optimisation process. However, the possible policies are reduced to the space of the piecewise linear functions with the selected number of points. If the data points are then required to have a pre-defined distance on the y-axis, the possible solutions are reduced further, but at the same time, the number of parameters and thus the number of free variables decreases. As in the previous case, the goal of the policy improvement is to find parameter values (i.e., data points of the table function), that improve the value of the objective function. A software package that supports table function policy design is found with the Powersim Studio plug-in SOPS (Moxnes and Krakenes 2005).

In both cases, the modeler has to define the functional dependencies of the policy function. This choice is closely related to the concept of bounded rationality (MoreCroft 1985; Simon 1984) models.

A policy function, i.e., a decision rule, is a model about what information cues an actor employs in order to make decisions in a given system. If this actor has only a limited view of the system, then the policy will only depend on the variables and information that are available to this particular actor (Sterman 2000). An improved policy will enable this actor to make better decisions based on the limited information available to him/her. Recent work has focused on improving policies for such actors, using, for instance, co-evolutionary analysis (Liu et al. 2012).

In this paper, we will consider a different kind of actor. Our actor has a global view of the model, i.e., he or she has information on all the state variables at all times within the simulation time horizon.

Modeling the policy of an actor with such a comprehensive level of awareness with the application of conventional approaches to policy analysis constitutes a difficult endeavor. One option would be to define a table function for each state, that depends only on that state. A mixed policy function that depends on all states, can then be defined as a sum of these functions (Keloharju and Wolstenholme 1989).

One conventional approach to System Dynamics optimisation is based on "optimisation by repeated simulation" (Liu et al. 2012). This has the advantage, that any model which can be simulated, can also be optimized, since there are no requirements on the properties of the model equations. However, approaches using repeated simulation suffer from the "curse of dimensionality" Bellman (2003) dynamic, where the significant dimension is that of the space of free variables. An additional free variable adds a dimension to the optimisation algorithm's search space. Solving optimisation problems with a large number of free variables therefore quickly becomes impractical. As a consequence, the degrees of freedom in a mixed policy function situation, are limited from a practical perspective, in the case of an optimisation of the policy by repeated simulation being attempted.

We present a different approach and in so doing, directly optimize the values of the policy function. This is equivalent to defining the policy as a time-dependent table function with one data point for each time step of the time horizon. In the context of physical systems, this kind of problem is known as an "optimal control problem" Betts (2011). With this approach no assumptions on the properties of the policy function are made *a priori*. It is only necessary to select the "free variables". In a conventional approach, these "free variables" would contain the values of the policy functions. For each of these variables, a range of valid values must be defined. It is then the task of the optimisation process, to find the optimal value for each free variable at each time.

The resulting optimisation problem based on a system dynamics model can be written as follows:

$$\begin{array}{ll} \max & c(x, y, z), \\ \text{s.t.} & \dot{x} = f(x, y, z), \\ & y = g(x, y, z), \\ & x(0) \in X_0 \end{array}$$

State variables: $x = x(t) \in \mathbb{R}^n$
Algebraic variables: $y = y(t) \in \mathbb{R}^m$,
Control variables: $z = z(t) \in \mathbb{R}^s$.

Time horizon: $t \in [t_i, t_f]$

In order to solve such a problem, we differentiate between two approaches:

3.2.1 Local Approach

In the local approach, the goal is to find a locally optimal solution. Local optimality means, that in a small neighborhood around the given solution, there is no solution with a better objective value. For this approach, standard methods exist for dynamical systems, which reliably deliver local solutions for small and moderately sized problems. The task at hand is to reformulate and adapt a system dynamics model, so that these methods can be used. Work on the local optimisation of system dynamics models can be found for instance in Vierhaus et al. (2014). In this chapter, we will focus only on the global approach.

3.2.2 Global Approach

In the global approach, the goal is to find a solution, and in addition to prove its global optimality. This means that no feasible solutions of the problem with a better objective function value exist. Hence, the global solution approach has two steps: Find an optimal solution and prove that no better solution exists.

Both of these approaches can prove successful using techniques from mathematical optimisation.

In the next section, we will show how modern optimisation techniques can be used in the global approach to system dynamics optimisation. The basis is the formulation of an optimisation problem, based on the control problem introduced in Sect. 3.2. As mentioned before, the simulation of a system dynamics model using numerical methods is well-established. This simulation is based on a time-discretisation of the model, which we will also use for our optimisation problems.

In order to discretise the model, we introduce a fixed time step of length Δt. We then consider the equations of (Sect. 3.2) no longer at any $t \in [0, T]$, but only at n_t points in time defined by $t = j \cdot \Delta t$, $j \in \{0, 1, \ldots, n_t - 1\}$. The derivatives appearing in (Sect. 3.2) need to be replaced by an appropriate discretisation scheme, for example a Runge-Kutta scheme. The resulting system can then be written as follows:

$$\max c(x_0, \ldots, x_{n_t-1}, y_0, \ldots, y_{n_t-1}, z_0, \ldots, z_{n_t-1}), \tag{5a}$$

$$\text{s.t.} x_{j+1} = f(x_j, y_j, z_j), \quad j \in 0, 1, \ldots, n_t - 2 \tag{5b}$$

$$y_j = g(x_j, y_j, z_j), \quad j \in 0, 1, \ldots, n_t - 1 \tag{5c}$$

$$x_0 \in X_0 \tag{5d}$$

$$\text{State Variables: } x_j \in \mathbb{R}^n \tag{5f}$$

$$\text{Algebraic Variables: } y_j \in \mathbb{R}^m, \tag{5g}$$

$$\text{Control Variables: } z_j \in \mathbb{R}^s. \tag{5h}$$

This system now has the standard form of an optimisation problem, similar to the one introduced in (1). In contrast to (1), we now only have a single objective function. On the other hand, we have nonlinear equality constraints in place of linear inequality constraints.

3.3 MINLP Approach

After the discretization of the system dynamics optimisation problem, it is possible to attempt to solve it with existing solvers. Since we are interested in global solutions, the algorithm used should be able to provide a certificate of global optimality. One group of solvers that can provide this certificate are the branch-and-cut solvers that were introduced in Sect. 2.3 This approach has been successfully applied in the solution of Mixed Integer Linear Programs as well as MINLPs from a range of applications [for example, see Defterli et al. (2011), Borndörfer et al. (2013), Humpola and Fügenschuh (2013)]. Solving a control problem derived from a discretised dynamical system with a standard branch-and-cut solver is, however, in many cases unsuccessful, since the solver does not take into account the special structure of the MINLP that arises from the discretization, and from the handling of non-smooth functions via integer variables. Without considering this structure, even finding a single feasible solution can exceed a reasonable time budget of several hours or even days.

In the remainder of this section, we will present the concept of a tailored solver for system dynamics optimisation problems. Like PolySCIP, this concept has been implemented in the framework of the modern MINLP solver SCIP and results can be found in Fügenschuh and Vierhaus (2013a, b), Vierhaus et al. (2014),

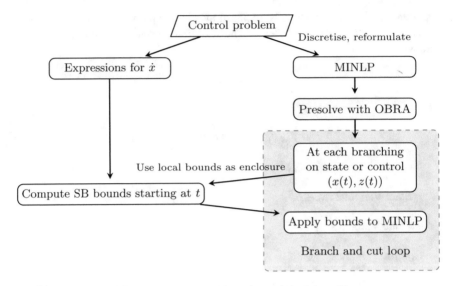

Fig. 7 Concept of a global solver for system dynamics optimization problems

Fügenschuh et al. (2013). A diagram describing the improved solution process is shown in Fig. 6.

3.3.1 Transcription

The first step is the reading and transcription of the system dynamics model and the optimisation parameters. Once the model and the optimisation parameters have been read, the optimisation model is processed in two ways. An equivalent MINLP is, then set up. This includes the time discretisation. At the same time, expressions for the function $\dot{x}(t)$ are derived from the model (Fig. 7).

3.3.2 Optimisation Based Reachability Analysis

To improve on the dual side of the algorithm, an Optimisation Based Reachability Analysis (OBRT) is performed for every problem. This analysis computes bounds for the possible states of the system using the dynamic behaviour and the initial values x_0 as input.

3.3.3 Primal Heuristic

In the interest of producing quickly feasible solutions, we implemented a simple heuristic that reduces the control problem to a simulation problem by fixing the control variables to their lower (or in a second run upper) bounds. If there are no path constraints, this process will always yield a feasible solution.

3.3.4 Bound Propagation Based on Differential Inequalities

To improve the bounds within the branch-and-cut process, we compute differential inequalities as outlined in Scott and Barton (2013). This involves the solution of an auxiliary simulation problem using the expressions for $\dot{x}$ derived in the reading of the problem.

3.3.5 System Dynamics SCIP

The concepts mentioned above have been implemented as the solver System Dynamics SCIP (SD-SCIP). Like polyscip, SD-SCIP is an extension of the modern MINLP solver SCIP and is publicly available (Füegenschuh and Vierhaus 2013a, b).

4 Conclusion

This chapter introduced the framework of multi-criteria optimization and system dynamics optimisation together with different modelling techniques. It showed that mathematical optimisation is a useful tool for modelling a wide variety of problems from the sustainability context. The two solvers presented PolySCIP (Schenker et al. 2016) and SD-SCIP (Fuegenschuh and Vierhaus 2013a, b) were specifically developed with applications from sustainability in mind. They can be used as decision support instruments for a wide range of problems, from scheduling, manufacturing and production to planning subsidies and taxes and exploring dynamical pathways into the future. Both tools are publicly available and present an opportunity for the sustainability community to benefit from recent advances in mathematical optimisation.

References

Achterberg, T. 2009. Scip: Solving constraint integer programs. *Mathematical Programming Computation* 1(1): 1–41. http://mpc.zib.de/index.php/MPC/article/view/4.
Bellman, R. 2003. *Dynamic programming*. Dover Books on Computer Science Series. Dover Publications.
Betts, J.T. 2010. *Practical methods for optimal control using nonlinear programming. Advances in design and control*. Philadelphia, PA: Society for Industrial and Applied Mathematics.

Bixby, R.E. 2002. Solving real-world linear programs: A decade and more of progress. *Operations Research* 50: 3–15.

Borndörfer, R., A. Fügenschuh, T. Klug, T. Schang, T. Schlechte, and H. Schülldorf. 2013. The freight train routing problem. Technical report, ZIB Technical Report ZR-13-36.

CPLEX. 2016. IBM ILOG CPLEX Version 12.6.3. www-03.ibm.com/software/products/en/ibmilogcpleoptistud.

Csirmaz, L. 2016. Inner. https://github.com/csirmaz/inner.

Dangerfield, B., and C. Roberts. 1996. An overview of strategy and tactics in system dynamics optimization. *Journal of the Operational Research Society* 47: 405–423.

Dantzig, G. 1963. *Linear programming and extensions*. Rand Corporation Research Study. Princeton University Press.

Defterli, O., A. Fügenschuh, and G.-W. Weber. 2011. Modern tools for the time-discrete dynamics and optimization of gene-environment networks. *Communications in Nonlinear Science and Numerical Simulation* 16(12): 4768–4779.

Ehrgott, M. 2005. *Multicriteria optimization*, 2nd ed. Berlin: Springer.

Fügenschuh, A., and I. Vierhaus. 2013. A global approach to the optimal control of system dynamics models.

Forrester, J.W. 1961. *Industrial dynamics*. Waltham, MA: Pegasus Communications.

Fügenschuh, A., S.N. Grösser, and I. Vierhaus. 2013. A global approach to the control of an industry structure system dynamics model. Technical Report 13-67, ZIB, Takustr.7, 14195 Berlin.

Fügenschuh, A., and I. Vierhaus. 2013. A global approach to the optimal control of system dynamics models. Technical report, ZIB Technical Report ZR-13-28.

Gurobi. 2016. GUROBI Optimization Version 6.5. www.gurobi.com.

Humpola, J., and A. Fügenschuh. 2013. A unified view on relaxations for a nonlinear network flow problem. Technical report, ZIB Technical Report ZR-13-31.

Keloharju, R., and E. Wolstenholme. 1988. The basic concepts of system dynamics optimization. *Systemic Practice and Action Research* 1(1): 65–86.

Keloharju, R., and E. Wolstenholme. 1989. A case study in system dynamics optimization. *Journal of the Operational Research Society* 40(3): 221–230.

Koch, T. 2004. *Rapid mathematical prototyping*. Ph.D. thesis, Technische Universität, Berlin.

Kutta, W. 1901. *Beitrag zur näherungsweisen Integration totaler Differentialgleichungen*. B.G Teubner.

Ladanyi, L., T. Ralphs, M. Guzelsoy, and A. Mahajan. 2016. SYMPHONY 5.6.14. https://projects.coin-or.org/SYMPHONY.

Liu, H., E. Howley, and J. Duggan. 2012. Co-evolutionary analysis: A policy exploration method for system dynamics models. *System Dynamics Review* 28(4): 361–369.

Löhne, A., and B. Weißing. 2014. Bensolve 2. http://www.bensolve.org.

Morecroft, J.D.W. 1985. Rationality in the analysis of behavioral simulation models. *Management Science* 31(7): 900–916.

Moxnes, E., and A. Krakenes. 2005. SOPS—a tool to find optimal policies in stochastic dynamic systems. In *Proceedings of the 23rd International Conference of the System Dynamics Society*, ed. by J.D. Sterman, N.P. Repenning, R.S. Langer, J.I. Rowe, and J.M. Yanni.

MPS Format. 2016. Mps format—Wikipedia, the free encyclopedia. Accessed 28 July 2016.

Richardson, G.P., and A.L. Pugh. 1981. *Introduction to system dynamics modeling with dynamo*. Cambridge, MA: MIT Press.

Runge, C. 1895. Ueber die numerische Auflösung von Differentialgleichungen. *Mathematische Annalen* 46: 167–78.

Schenker, S., R. Borndörfer, M. Skutella, and T. Strunk. 2016. PolySCIP. In *Mathematical Software –ICMS 2016, 5th International Congress, Proceedings*, ed. by G.-M. Greuel, T. Koch, P. Paule, and A. Sommese, vol 9725 of *LNCS*, Berlin, Germany. Springer. http://polyscip.zib.de.

Scott, J.K., and P.I. Barton. 2013. Bounds on the reachable sets of nonlinear control systems. *Automatica* 49(1): 93–100.

Simon, H.A. et al. (1984). *Models of bounded rationality*, volume 1: economic analysis and public policy. MIT Press Books 1.

Sterman, J.D. 2000. *Business dynamics—systems thinking and modeling for a complex world.* Boston: Irwing McGraw-Hill.

Vierhaus, I., A. Fügenschuh, R.L. Gottwald, and S. Groesser. 2014. Modern nonlinear optimization techniques for an optimal control of system dynamics models. In *Proceedings of The 32nd International System Dynamics Conference.*

Xpress. 2016. FICO Xpress Optimization Suite Version 7.9. www.fico.com/en/products/fico-xpress-optimization-suite.

Yücel, G., and Y. Barlas. 2011. Automated parameter specification in dynamic feedback models based on behavior pattern features. *System Dynamics Review* 27(2): 195–215.

Inducing Behavioural Change in Society Through Communication and Education in Sustainable Manufacturing

Ina Roeder, Wei Min Wang and Bernd Muschard

Abstract The United Nations considers the mobilization of the broad public to be the essential requirement for achieving a shift towards a more sustainable development. Science can play a vital role in Education for Sustainable Development (ESD) by contributing to ESD-related research and development on the one hand, and by becoming active awareness raisers themselves in education and multiplier networks. Specifically, the use of special *Learnstruments*, and investment in *Open Education* formats among other educational tools, may pave the way for accelerated apprehension and appreciation of sustainable manufacturing topics among the greater populace.

1 The Challenge of Creating Proper Understanding of Sustainable Manufacturing

For all liveable future scenarios, a change of manufacturing paradigms is mandatory, not only by producers but also by customers and users. In order to realize such a behavioural change in society, it is essential to establish proper appreciation of sustainable manufacturing or in a broader perception the general concept of sustainable development. One conceptualization of a learning process holds that people have to acquire knowledge and interpret and apply it to their own personal contexts (Kolb 1984; Kirkpatrick 1996) in order to learn the lessons at hand. To assist people in undergoing this learning process, awareness of sustainable development has to be

I. Roeder (✉) · W.M. Wang · B. Muschard
Institute for Machine Tools and Factory Management,
Technische Universität Berlin, Berlin, Germany
e-mail: ina.roeder@tu-berlin.de

© The Author(s) 2017
R. Stark et al. (eds.), *Sustainable Manufacturing*, Sustainable Production,
Life Cycle Engineering and Management, DOI 10.1007/978-3-319-48514-0_16

255

raised first and foremost, and the respective knowledge has to be disseminated accordingly. A range of factors however stands in the way of that pursuit.

Firstly, the complexity attached to the concept of sustainable development impedes distinct understanding. It is often criticized as missing clear outlines and being applied inconsistently (Grunenberg and Kuckertz 2005; Michelsen 2005; Brand 2005). The predominant sustainable development model used today entails the three pillars or spheres of sustainability, which emerged with the United Nations Report "Our Common Future" by Harlem Brundtland in 1987. This model states that sustainable development is only possible when all three spheres—economic, social and environmental—are equally addressed. It was this attempt of a super-framing that successfully combined the diverse perspectives and claims that competed for leadership within the sustainability discourse in the beginning of the 1990s (Brand 2005). It was a concept that everyone could agree upon, as it was broad enough to contain contrary perspectives. The other side of the coin is that such a concept is inevitably inconsistent and therefore lacks clear outlines. From a layperson's viewpoint, this concept leads to contradictory scenarios, wherein singular measures serve to increase sustainable development and reduce it at the same time, e.g. when a turn towards environmentally friendly products and more selective consumption patterns leads to job cuts, unemployment and higher poverty rates at the production site.

Secondly, the popular spin of the term fails to mobilize people. As of the 1990s, the public debate that later turned into sustainability communication still had a clear environmental framing. Fuelled by catastrophes such as in Bhopal (*1984*) and Chernobyl (*1986*), with strong media coverage, environmentalism became a social representation, an element ultimately endowing social groups with identity (Kruse 2005). Consequences were political activism, broad framing in educational institutions, the media and the private sphere alike, and a sheer explosion of well-designed information. In short, it triggered strong reactions in civil society and central tenets which were fully embraced into people's thinking. Yet the phenomenon did not get repeated when the debate turned from environmentalism to sustainable development in the aftermath of the United Nations Conference on Environment and Development in Rio de Janeiro, 1992. In this case, social and economic concerns were added to the agenda of environmental threats (Michelsen 2005). However, this did not translate into an increase in private activism nor into the internalization of higher urgency due to heightened threats to societal welfare. On the contrary, when the concept of sustainable development as a multi-perspective issue was introduced, a strong trend of "de-dramatization" (Grunenberg and Kuckertz 2005) set in, which persistently increased in the following decade. The challenges and possible measures were communicated and regarded as less immediate and rather long-term in their effects, which resulted in lower level short-term mobilization.

Consequently, despite society's increasing familiarity with the sustainability terminology, appreciation of the overall concept and awareness of its concrete meaning in everyday life remain low. In Germany, for instance, 15 years of intensive efforts to communicate sustainability through federal institutions and

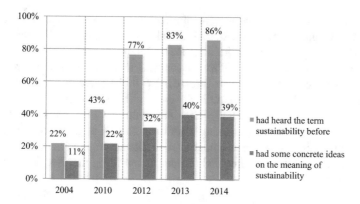

Fig. 1 Average Germans' acquaintance with the term sustainability over time (Roeder et al. 2015)

broad media coverage endured with some effect on people's awareness of the topic as shown in Fig. 1 (Roeder et al. 2015). Still, in 2014 only 39 % of the people had some concrete ideas on the meaning of sustainability and less than about 4 % associated it with future-aware behaviour. As these facts apply to Germany, a nation known to have an elaborated educational system and easy universal access to information, the direness of the information campaign can be expected to apply even more seriously to people from parts of the world with little access to information and a low level of basic education.

With the sustainability challenge becoming increasingly urgent, awareness training continues to be a central task of all activities aiming at sustainable development. This holds especially true for the field of sustainable manufacturing, which is so far widely neglected in public discourse, in spite of its great impact on all areas of human living. To be sure, the educational frameworks for school education have been recently rewritten in Germany to incorporate sustainable development into the curricula as a basic principle as well as a specific learning objective. Nevertheless, the sustainability impact of manufacturing is hardly considered (Roeder et al. 2016). However, considering the German example described before, classic measures seem to have failed so far in communicating the complexity of sustainable development, and especially sustainable manufacturing, to people with little previous knowledge. Just as sustainable development can only be achieved when activating the majority of populace, this majority can only be won over when stakeholders from diverse fields of sustainable manufacturing are activated to join in and strengthen change in society.

This chapter is meant as a guide to support the planning of knowledge dissemination measures in multi-disciplinary research projects. A general approach for sustainability communication is introduced to highlight integral aspects in the planning process. Furthermore, present gaps regarding mediation of knowledge about sustainable manufacturing are identified. By providing best-practice examples, it will be demonstrated how specific challenges can be met. A central aspect addressed in this context is Education for Sustainable Development (ESD), which aims at teaching competencies as a combination of certain skills and knowledge that enable the learner to understand, judge and act according to the sustainability maxim (Wals 2015). Education for Sustainable Manufacturing (ESM) in this regard is seen as a partial aspect of ESD with concrete focus on industrial aspects. The importance of education is also stressed by the Organisation for Economic Co-operation and Development (OECD) and the United Nations (UN), as, both organizations agree on education being the main resource for societal change towards sustainable decision-making (Bormann 2005). It is further argued that science, in its unique position as a neutral and reliable source of knowledge, should figure into the equation as a key stakeholder in spreading the word of sustainable manufacturing.

2 General Approach for Science-to-Public Sustainability Communication

As sustainability communication intends to reach a great number of people, it can be considered as a form of mass communication. Following the fundamental model of mass communication developed by Lasswell (1948), every action in this context should be designed by asking the "five Ws": who says what, in what way, to whom, and with what effect? Although widely criticized for its ignorance of the receivers' active role in influencing the communication by giving feedback to the sender of the message, those "five Ws" represent, to this day, the major fields of mass communication science. Answering these questions in the context of sustainability communication from a scientific point of view, forms the boundary conditions for the respective communication framework.

"Who"—The Communicator

The role of communicators in their domain and their intended communication goal, imparts a strong influence on the message, the channels and the target groups. This matter of *who* does the communicating is also key to where the problem lies. Communicator credibility depends on status and expertise on the one hand and on affectionately ascribed trustworthiness on the other. For the US it has been shown that professors are ascribed both, expertise and trustworthiness (Fiske and Dupree 2014). This gives them an excellent initial position as communicators for people will tend to believe them and agree with their opinions. Contrarily, scientists, researchers and engineers are seen as experts but tend to be allocated less trust, which reduces their credibility ascribed by the broad public. However, people's trust in someone changes significantly with this person's position in relation to the position of those who judge. This means, while the majority may not ascribe great trustworthiness to scientists, researchers and engineers, the result is different when asking sections of society that have certain aspects in common with those communicators, e.g. a high educational level. Also the ascribed trustworthiness is expected to increase when those scientific communicators show concern for humanity and the environment; both being the case for manufacturing-oriented sustainability communication.

For people transmitting sustainability knowledge, such as teachers, the greatest capital is knowledge. These educators need to be sceptical towards new information which they are persuaded to implement in their teaching activities by non-official bodies, and, furthermore, be concerned, among other things, about the correctness of the information and the underlying interests of the persuader. This locates them near science communicators, making them a convenient target group for science communication. When it comes to decision-makers (e.g. in politics), the reputation of an information provider who is well-established in a certain field of expertise, offers opportunities with influential stakeholders and increases the chances of being heard. This is where publically funded science has an invaluable advantage. It is considered neutral and exact in the highly competitive arena of sustainable manufacturing.

As a communicator, science has a vital position in passing on knowledge. Hence, it has a triple role to play in (1) generating communicable knowledge about sustainable manufacturing, (2) developing new scientifically sound dissemination techniques and acting as a communicator with great credibility, and (3) promoting knowledge dissemination and awareness raising for sustainable manufacturing. Consequently, communication and teaching aspects should be considered in every research project within the field, right from the very planning phase onwards.

"What"—The Message

The overall message of sustainable development is clear—we need to live in such a way that future generations can have an average standard of living which is at very least equal to the one we have today. The message of sustainable manufacturing is even more narrowly defined, insofar as stating that dynamics of global competition and cooperation can be used for lending wings to processes of innovation and mediation towards the goal of global sustainability. Clear as those definitions might appear in this abstract form, thorough understanding of the concepts requires profound understanding and perspectives that are currently lacking in the narration of the public discourse and thus hardly intuitive. To enable knowledge of sustainable development and manufacturing, and to facilitate that message getting communicated in a comprehensive way, it has to be applied to the context of the target groups, e.g. by relating it to monetary values for industrial producers or strategic advice in daily life situations for consumers. As shown in this book, a multitude of examples demonstrate how technological, social and economic innovations can be integrated with each other to contribute to sustainable development by means of saving resources, increasing the living standard throughout the world without increasing consumption, and developing business models that are based on functionality rather than on personal ownership. Particularly with regards to communication to the broad populace, a crucial aspect of the message is to raise awareness about the complex nature of sustainability. The goal should be to create a differentiated understanding of the term and hence to allow for sophisticated decision-making in daily life.

"To Whom"—The Target Group

Considering the communication goal of changing people's behaviour, and the findings on credibility described above, it becomes obvious that it is insufficient to simply view the broad public as one homogeneous target group. Moreover, experience from former sustainability communication measures shows us that mass coverage can only play a supportive role in the whole process (Roeder et al. 2015). With respect to the variety of potential recipients and communication goals, no panacea exists. Hence, addressing multipliers becomes an integral part of mediating knowledge to large numbers of diverse recipients. Multipliers can be defined as persons who have the ability to influence the opinion, the behaviour or the actions of a social group by virtue of the authority assigned e.g. by their social status or professional expertise. Their relevance results from their hybrid nature, as they constitute just as much the target group as they do the role of communicator. Multipliers can be, for example, teachers, trainers or any other people in positions who communicate with a great number of citizenry in their day-to-day work. They can also be decision-makers who influence a lot of people's behaviour by deciding on the choices they get to make, e.g. product designers or politicians. Lucky for scientifically-based sustainability communication, those are the very target groups who are likely to ascribe publicly funded science communicators high credibility, as argued above.

By involving multipliers as a mediating party, a simplified model of sustainability communication has been introduced that consists of three sets of communicators and target groups respectively. All three parties together represent the communication network of science-to-public sustainability communication (Fig. 2).

Each party has to be understood as a communication partner who possesses valuable information on sustainable development and power to influence its dissemination into society. For instance, teachers can give information on what materials or tools they require for teaching sustainability. Decision-makers have insights into the constraints that influence people's behaviour, which often go unnoticed. The

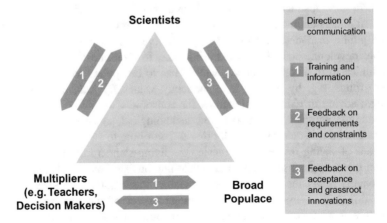

Fig. 2 Simplified communication model for dissemination of sustainability knowledge from a scientific stakeholder perspective

broad populace may have information about the acceptance of sustainability measures as well as about grassroots innovations and movements.

Viewing education as a core vehicle for transferring sustainability knowledge into society allows for a more differentiated view of the target groups. While the OECD and the UN consider education at all levels of formal and non-formal education, teaching a holistic understanding of sustainable manufacturing requires more specific target groups. Although it is useful if general ideas of sustainable development are taught from early childhood onwards in conjunction with a uniform set of values, the integration of industrial aspects such as technology, production planning and business models, should wait until the learners' cognitive ability has matured enough to process such complexity.

The human brain develops rapidly up to the age of about twelve. At the age of 13, further increase in memory performance is usually slow and marginal (Ahnert 2014). The ability of hypothetical and scientific thinking emerges, enabling the young learner to verify hypothesizes by using logic. The cognitive ability developed by adolescence enables the students to rapidly extend their semantic networks from that point on (Ahnert 2014). Well-developed semantic networks are fundamental to complex thinking, such as needed for understanding the workings of sustainable manufacturing challenges and solutions. It can be therefore clearly recommended to concentrate on target groups from the age of around 13 onwards when teaching complex aspects of sustainable manufacturing. Of course, it helps by

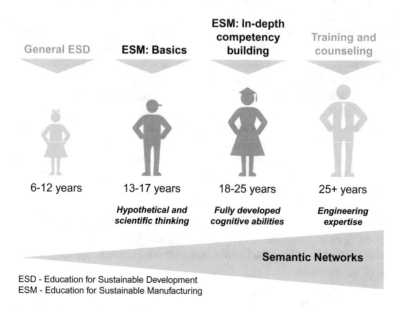

Fig. 3 Levels of education for manufacturing-related sustainable development

all means if the students are already familiar with more general aspects of sustainable development and science by that time, as demonstrated in Fig. 3.

Teaching sustainable manufacturing at the high school level again lays the foundation for easy integration of correlating assumptions into higher education. Still, the main focus of engineering and engineering economics education lies in classic paradigms such as profit maximization. Sustainability aspects are inadequately represented despite the dire need to sensitize future manufacturing experts to their responsibility as decision-makers and teach them how to plan and implement sustainable manufacturing. In summary, ESM, although generally building upon ESD, needs to target high school students in order to prepare them for further training as engineering or engineering economics students in higher education so to pave the way for new, sustainability-oriented paradigms in manufacturing. Also targeting the youth in general means targeting the next generation of consumers, whose product choices make them direct stakeholders of sustainable manufacturing if they choose to invest in sustainable products and sustainable production. Through the same mechanism they can also have indirect effects as a pressure group on enterprises that still follow unsustainable manufacturing strategies.

"In What Way"—The Channel

Target group orientation is the core of successful communication. The channels that are used are therefore as manifold as the target groups to be communicated with. Those channels can be direct or indirect, depending on the assignment of the target group as shown in Fig. 4. Apart from research-based communication such as interviews in direct communication and survey sheets in indirect communication, the focus of direct communication with multipliers is on training and through active participation on the part of the respective stakeholder networks. The broad populace can best be reached by offering exciting events with a high entertainment factor or even public educational projects. Indirect communication can work by offering specific training materials such as extended teacher manuals complete with teaching materials or materials for qualifying teachers as "Teachers of ESD" as a labelled skill enhancement, for example. Training materials and appropriate manuals for skill-enhancement likewise play a major role in the indirect communication with multiplying decision-makers, especially from industry. Broad populace is thus reached indirectly through teaching or through informational materials offered by the trained multipliers, and also through a variety of activities such as exhibitions or competitions.

Useful communication formats and tools differ greatly among the target groups. It is necessary for effective communication to choose carefully the channels that are to be used. The channels described above are meant to be supplementary to the well-established channels of scientific and journalistic media production such as articles or print media.

"With What Effect"—The Result

Just as the impact of every communication activity should be measured and every new product should be tested, the impact of innovative ESD activities needs to be monitored in order to identify undesirable effects or outright ineffectiveness. The

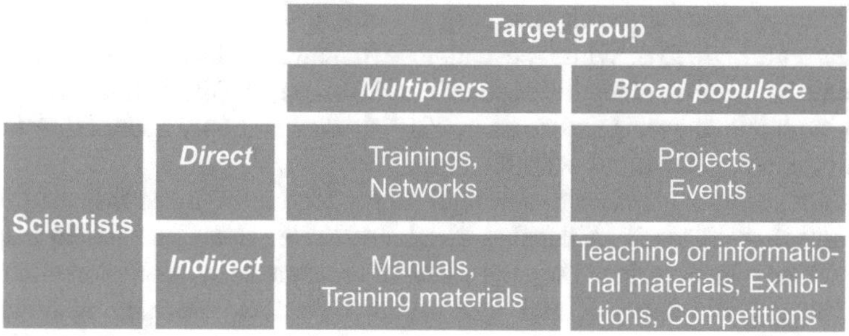

Target group		
	Multipliers	*Broad populace*
Scientists **Direct**	Trainings, Networks	Projects, Events
Indirect	Manuals, Training materials	Teaching or informational materials, Exhibitions, Competitions

Fig. 4 Exemplary ESD communication from a scientific stakeholder perspective

outcome of studies on knowledge gained and attitudinal or behavioural change can usually not be expected to represent a fixed reality. It lies within the nature of social sciences that there are as many social realities for a surveyed person as there are social or psychological circumstances which this person experiences. The situation becomes even more complex when the participants are children whose semantic webs and other cognitional modes are not yet fully established (cf. Ahnert 2014). In that vein, planning research designs for such target groups proves to be challenging. Pre-tests of the design are thus absolutely necessary in this context. Especially if a research group's main focus lies in the technological field—as to be expected when it comes to sustainable manufacturing—social scientific expertise needs to be integrated in order to confront this challenge. However, a great number of cases and careful research design can provide valid data on knowledge, attitudinal and behavioural development subsequent to a treatment e.g. an ESD measure. This data is fundamental to developing effective ESD solutions that are capable of contributing to the societal change of paradigms towards sustainable development.

3 Present Gaps and Best Practice Solution Examples

This section presents exemplary gaps in ESD and ESM which were identified in the course of an interdisciplinary research project on sustainable manufacturing. In the following paragraphs, some of these gaps are introduced in context, along with best practice solutions.

3.1 Sustainable Manufacturing in High School Education

A special focus of the Agenda 21, the UN development program for the 21st century, lies with children and teenagers. In Germany, the programs "21" and "Transfer-21" have been set up as local forms of the Agenda 21 from 1999–2008 in order to improve sustainability teaching at German schools, with moderate success (Roeder et al. 2015). While educational frameworks have been rewritten in Germany in order to integrate sustainable development into formal education, a survey with above-average students in 2014 showed that only about 50 % had any future-oriented associations with the term.

In-depth sample interviews with high school teachers showed that they did not feel competent to teach sustainable development, let alone sustainable manufacturing (Roeder et al. 2016). They felt a lack of fundamental appreciation of the topic of sustainable development and furthermore lacked the teaching materials that would help them to overcome their knowledge deficiency in class. That this notion is a common one among teachers becomes apparent in a study with educators from schools that are implementing ESD programs under a local German program in 2015. Although all participants are already involved in ESD activities and have been offered qualification courses, 44 % say it is difficult to develop the necessary competencies for teaching ESD, and 51 % claim, moreover, that it is difficult to find adequate teaching materials.

3.1.1   Open Educational Resources

The challenge of lacking adequate teaching materials for a fast developing field with multiple perspectives could be met by solutions from the open knowledge movement. That is, high expectations for educating the populace worldwide have been raised by the concept of so-called Open Educational Resources (OER). The Paris Declaration of the UNESCO 2012 World Open Educational Resources Congress defines OER as "any type of educational materials in the public domain, or released with an open license, that allows users to legally and freely use, copy, adapt, and re-share".

OER are dynamic. They can be quickly adapted and shared since they are supposed to be produced in an open format and shared online. They also allow for a wider variety of cases and examples than can be covered by a textbook alone. OER thereby encourage teachers to tailor their teaching units according to their students' interests or current debates. This is where topics such as sustainable manufacturing, which are widely neglected in education so far, can still be brought to teachers' attention.

Sustainable development is mainly scheduled for the 9th and 10th grade at German high schools (Roeder et al. 2016). A search for German OER on sustainable development linked with topics of technology or industry for this target group in 2015 brought 29 results of which 18 also included at least one working sheet to use in class. Most of them had been developed for the subjects of geography, social sciences, biology, politics, religion/ethics, and economics. An analysis using the LORI[1] method, assessing the items in seven categories on a 5-point scale with 5 being the maximum score, showed an average (arithmetic) score of 3.6. Although some

---

[1]Learning Object Review Instrument by Leacock and Nesbit (2007).

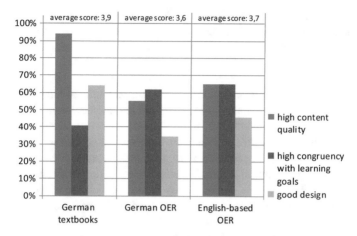

Fig. 5 Comparison of teaching materials assessed with the LORI method

resources, especially those from official bodies, scored very high, only 55 % had good (4) or very good (5) results at the assessment of content quality as shown in Fig. 5. Another weak spot has been identified to be design: Only 10 out of 29 items scored good to very good. 62 % had high or very high congruency with the defined learning goals and 63 % included motivational elements such as varying assessment types. Generally the OER scored lower than the sustainability sections of geography text books for the same target group; those having an average (arithmetic) score of 3,9.

The exemplary international search for English OER on sustainability and technology or sustainability and industry for the same target group (n = 48, 23 including working sheets, 131 identified items total) revealed the USA and Canada to be the main producers of OER on the topic for this specific target group. However, there are also free English teaching materials accessible by providers from the UK, Australia, Norway, and France among others. The LORI assessment of 48 items that met the requirements of topic and target group best showed a slightly higher score of the English OER than of those produced in German language, the average (arithmetic) score of the international OER being 3,7. 65 % of the English-based OER were assessed to have good or very good content quality and congruency with the learning goals. 22 out of 48 assessed items scored good to very good with regard to design.

Apart from often poor didactic design, the connection to core sustainable manufacturing topics were only marginal in most cases. This is a gap that needs to be filled if the topic stands a chance of getting incorporated into high school curricula. Since content quality is one of the weak spots, science has a clear advantage as a producer of up-to-date and technically sound content. Critical in that pursuit is that the research teams intending to produce OER as a tool for raising awareness for sustainability must be multidisciplinary and bring together technical and didactic expertise. An example of this has been done within the Collaborative Research Centre (CRC) 1026 "Sustainable Manufacturing—Shaping Global Value

Creation" (Roeder et al. 2016). When developing and producing their teaching unit on "Sustainable Manufacturing," the scientists followed a 3-step action plan covering content definition (1); didactic structuring (2); and material production (3).

Content Definition

Resource consumption in manufacturing is the central theme of the OER developed by CRC 1026, addressing matters of human, natural and economic resources. In a first teaching unit, general information on sustainable development built up to the connection with manufacturing issues, so that, for example, the three pillars of sustainability were explained from a manufacturing perspective. This was then exemplified by a second unit discussing bicycle production in the context of more specific sustainability issues within global value creation, such as producing in low-wage countries, distributed production and CO_2 emissions. A third unit addressed Maintenance, Repair and Overhaul as clearly technical topics of sustainable manufacturing, also addressing, for example, planned obsolescence.

Didactic Structuring

In the interest of implementing the educational frameworks with the ultimate criteria of introducing teacher materials, nationwide educational programs were analysed for their explicit reference to sustainable manufacturing. It became clear that sustainable development is mostly set to become a fixed part of the 9th and 10th grade curricula and for all geography classes in nearly all federal states. To that end, the content was defined according to these frameworks' competencies and learning goals, such as "cosmopolitan acquisition of knowledge, including multiple perspectives" which was met, for example, by means of a role-playing exercise in which students take on various roles of producers, workers and customers from different geographical and cultural backgrounds. The learning goals of each exercise and their links to the educational framework, along with further didactic information, were all made explicit in an accompanying teacher's guide.

Each unit was structured following a reduced learning spiral oriented at Mattes (2011). To sum up, the procedure starts with teacher-oriented learning, requiring increasing self-study and group study as the lessons proceed, and finally ending with teacher-led concluding elements which follow up on individual learning results. Obviously the content must be general in the beginning, using everyday experiences of the target group as the starting point. It gets more specific as the lesson proceeds. At the end of the lesson, exercises are designed to ask students to transfer the acquired principles to other fields.

Since the material is supposed to be usable at different proficiency levels, a focus has been set on internal differentiation. Hence, exercises are set in three levels of difficulty.

Material Production

The best content will be ignored by teachers and students alike if the design is not appealing. The CRC 1026 invested in a professional designer for layout and graphics. OER are free of charge and free to adapt. In using OER, it is however of utmost importance either only to use graphics that are offered under a global commons license, or to produce them explicitly as such. A challenge when creating OER is adaptability. A publishing licence allowing for adaption is no benefit if the format and design of the materials offered are themselves not adaptable. It is thus paramount that the designer does his/her work with software that most teachers or even students have access to. In that vein, CRC 1026 decided to do its layout in Microsoft Powerpoint in order to foster easy exchange of graphics or text blocks.

3.2 Sustainable Manufacturing in Higher and Vocational Education

Promoting excellence in engineering has emerged as a strategic goal on the part of industry, society and nations in pursuit of improving living standards. The European Technology Platform for Future Manufacturing Technologies (Manufuture) highlighted the role of engineering education explicitly as a key driver in achieving this goal (Manufuture 2006). Chryssolouri recommends "*manufacturing education should follow new approaches so as to prepare industry for the next-generation innovation and the support of its growth*" (Chryssolouris 2005).

Innovative sustainable manufacturing offers a vehicle for coping with the challenge of sustainability. New training and education activities within organizations comprise the lever for achieving higher education in this area. For structuring an engineering design course with respect to teaching aspects of sustainability, Pappa et al. (2013) took Bloom's taxonomy of the cognitive domain as a basis. Yet the development of an approach in engineering wherein instruments are used to convey aspects of sustainable manufacturing with regards to the affective and psychomotor domains, was however hardly discussed.

3.2.1 Learning Through the Support of Technology—*Learnstruments*

Great potential for increasing the awareness and the learning and teaching productivity on sustainable manufacturing topics is seen in addressing the matters of technical content and the learner's feeling, values or psychomotor skills at the same time. Such instruments for learning could be found in so-called *Learnstruments*.

Learnstruments are production technologic objects both tangible and intangible, automatically demonstrating their functionality to the user. They aim at increasing the learning and teaching productivity and expanding the awareness of the environmental, economic and social perspective of sustainability. By their application, Learnstruments enhance organizations' human, structural and relational capital through higher skills and knowledge, structure and collaboration.

The neologism Learnstrument consists of the words *learning* and *instrument*. Learnstruments support the learning process by providing adequate learning goals to the user. Instruments in this sense are considered as objects supporting the user effectively and efficiently in achieving the learning goals. Furthermore, learning processes can be designed in a new fashion, focusing on sustainability to shape people's understanding of this important topic during training and learning.

They address cognitive, affective and psychomotor learning goals and strive towards the fulfilment of high level learning goals. Enabled by new and existing information and communication technology, Learnstruments allow the determination of the user's cognitive learning level and provide adequate learning goals towards the fulfilment of creation. Repetition strengthens the user's psychomotor ability for adaptation of human skills to execute manufacturing tasks.

The concept of Learnstruments is introduced and illustrated with two prototypical implementations.

3.2.2   CubeFactory

The CubeFactory is a Learnstrument addressing the understanding of a closed loop material cycle of polymers by an application-oriented mediation process. This mini-factory constitutes self-sustaining learning and production equipment which contain the main components involved in value creation, such as material processing, energy supply, manufacturing tools and tools for knowledge transfer. Based on the learning cycle of Kolb (1984), the CubeFactory considers aspects of perception and processing continua designed to increase learning productivity. The user is methodically supported in knowledge creation by the elements of concrete experience, reflective observation, abstract conceptualization and active experimentation. An open source 3D printer is the main value creation tool. The additive manufacturing process is regarded as sustainable since it places material exactly where it is needed to build up the workpiece. Unlike subtractive processes such as turning, milling, drilling, virtually no waste or by-products are generated in the whole process.

The so-called Home Recycling Device (HRD) serves as a material supplier for 3D printer consumables and demonstrates the value and potential of plastic recycling. A mechanical knife-shredder granulates thermoplastic waste that is further processed into an electrically heated screw extruder. This can turn a non-valuable object like thermoplastic domestic waste, into a valuable product like 3D printer filament. "Comparing the cost of 100 kg of sorted plastic waste ($1.00) with 1 kg of 3D printer ABS-filament ($25), an up lift ratio of 2500:1 is realized" (Muschard and Seliger 2015; Reeves 2012). Through the application of the HRD, the user learns that local processing of raw materials can shorten or even eliminate distribution channels, can reduce the volume of waste, can save on CO_2 emissions, and at the same time ultimately make the production of goods more cost effective. An important lesson in the mediation of sustainability is that energy cannot be produced, but only converted. In a sustainable manner, it applies to abdicating non-renewable resources and to making renewable resources available.

For those purposes, the CubeFactory contains a self-sufficient energy supply system formed by solar modules, rechargeable batteries and a battery management system. The knowledge transfer device is a learning environment implemented in a touchscreen tablet computer, supporting the user in exploiting the potential of the mini-factory. It assists the user in comprehending the CubeFactory's manner and in carrying out learning tasks in a simple and intuitive way.

To address a broad spectrum of users, to arouse curiosity and to motivate the learner, the CubeFactory is designed taking differences in knowledge, skills, age, disability or technological diversity into account (Fig. 6).

Fig. 6 CubeFactory: mobile, self-sufficient mini-factory

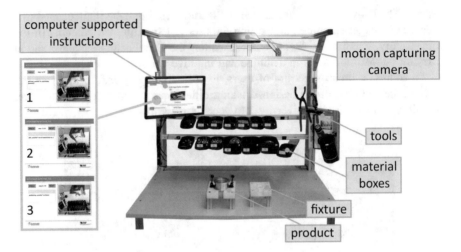

Fig. 7 Smart Assembly Workplace: assembly sequence of bicycle e-hubs is automatically transmitted to the user

3.2.3 Smart Assembly Workplace

The Smart Assembly Workplace (SAW), shown in Fig. 7, is a learning workplace for manual (dis-)assembly tasks with the example of bicycle e-hubs. It equips the worker with the tools and know-how needed to improve and plan such a workplace on their own. The learning-path is structured in initial learning and consecutive in-depth e-learning. It consists of fixtures, material boxes, tool holders and a camera to be affixed at the workplace.

During initial learning, users less experienced in assembly obtain a basic overview of the assembly sequence. The main requirement for this is to give the user immediate feedback referring to her/his current constitution and actions. By means of a marker-less motion-capturing software (Krüger and Nguyen 2015), the hands of the user are tracked by the system. Whenever the learner enters a so-called event-zone, an internal time stamp is logged and the assembly description automatically reveals the next assembly step on the display. In case of a mistaken action, a message is displayed to the user.

When the user enters, for example, the nuts-bunker with her/his hand, it can be assumed that at least one nut has been picked. On the basis of the time spent, conclusions with respect to the current work performance or level of learning of the user can be drawn. As soon as the worker's performance reaches the target time according to Methods-Time Measurement (MTM), the respective MTM-code is displayed to the user via the computer-supported instruction. It is utilised for the purposes of analysis and planning of working systems. By this representation, the user implicitly learns about the composition and meaning of the respective code.

The learner can use an e-learning module facilitating MTM knowledge in a self-explanatory way. The module consists of descriptions, hints and recommendations about the usage of MTM with the example of the bicycle e-hub. In a final stage, generic suggestions for improvement are displayed to the learner. These improvements are dedicated to assisting in the process of creating ideas for improvements in the learner's workplace (McFarland et al. 2013).

Although learning and understanding are intrinsic processes, this happens mostly in the setting of an interaction between the learner and the environment. Intelligently designed technologies and artefacts can assist the human in her/his learning process, and help to enhance teaching and learning productivity. The increasing digitization of manufacturing opens up new opportunities for knowledge transfer, in which the teacher and the learner no longer need be present at the same location.

The SAW replicates the production technology laboratory of the Vietnamese-German-University in Ho Chi Minh City, Vietnam. An assembly description, recorded at the German SAW, was transferred to the Vietnamese one. It was shown that the students in Vietnam—having scant knowledge about assembly —were able to assemble e-hubs with the help of this description. An expert was not required to be present in Vietnam to that end at all.

3.3 Facilitating Appreciation of Sustainability Aspects Through Gamification

As described above, the topic of sustainability is rather complex and it therefore takes time to supply an interested person with the necessary knowledge. In the context of the general public, the interest in picking up information without being forced to (by work, school or similar) decreases if too much time is required to supply the knowledge. Gamification addresses this topic by the use of game design elements in non-game contexts (Tan et al. 2011). Gamification provides elements that keep the interest of a person in a specific topic by using design elements like scores, achievements and storylines.

One way to transfer and demonstrate the challenge of sustainable product development is to let people experience this process first hand. Therefore, a "Product Configuration Game" (PCG) was developed in which the user is put in the role of a product developer who has to configure a new product from a limited set of options (Wang et al. 2014). The product in that case is a simplified model of a so called Pedelec (Pedal-Electric Bicycle). The configurable parts of the Pedelec comprise the basic frame and additional functional features. Furthermore, three different suppliers for the basic frame are available. This limited set of configuration options simulates existing supply chains and product politics. All features and product options are assigned with sustainability scores indicating their impact on respective sustainability indicators, such as global warming potential, primary energy consumption or fairness of salary. These scores where derived from results from a LCA conducted by Neugebauer et al. (2013) for a similar use case. By aggregating all sustainability score of one specific setting, a total sustainability score is calculated and visualized as bar chart for each of the sustainability dimensions (see Fig. 8). To demonstrate the fact that product developers usually do not have all necessary information about the impact of their choices the visualization of the total sustainability impact is also not available at the beginning of the game. Instead, the users have to rely on vague descriptive characteristics of features, such as material price, weight or design style. Only when they confirmed their decisions the bar charts representing the sustainability impacts are revealed. Then the users can change their decisions to explore the influence of different options. The impacts of their changes are then shown in real time. A further PCG feature, called the "Ontology Browser" allows the user to investigate the complex network of relationships between the product options and the sustainability indicators in a controlled way by using ontological trees developed for this game (Wang et al. 2014).

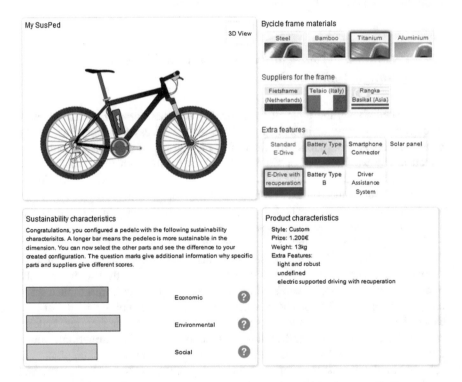

Fig. 8 Product Configuration Game: the user interface provides graphical feedback in the product model and shows impact of configuration decisions on all three sustainability dimensions in real time as bar charts

Various Gamification Design Elements (Tan et al. 2011) where chosen to motivate the user:

- Mechanics of the configurator construct a system of interacting parts that can be combined to achieve different results, so that exploring the different types of sustainability impact of the pedelec parts is necessary in order to understand the game mechanics
- Feedback visualization shows the result of the combination by delivering not only values in terms of graphs but also by providing a visual of them using a 2D/3D representation of a pedelec, therein enabling one to create her/his own custom-designed bike
- Fun motivator—role-play puts the user into the role of a design engineer with the task of creating a sustainable pedelec
- Fun motivator—research uses the ontological mechanisms for providing a visualization of the complex network behind the sustainability of the pedelec, which then allows the user to explore those networks discovering new relations

Using these gamification elements enriches the configurator in a way that users are kept interested as they are supplied with more information about sustainability during the usage of the configurator.

4 Conclusion

If the lifestyles of both economically up-coming and economically developed communities are persist to be shaped by the existing, currently predominant technologies, then resource consumption will exceed every accountable ecological, environmental and social boundary known to man (Seliger 2012; Ueda et al. 2009). However, human initiative and creativity opens up a panoply of paths for future development in pursuit of coping with the challenges of sustainability on a globe scale. Their chances of successful implementation essentially depend on their ability to take hold in an increasingly globalized arena of market driven activities. Both, demand and supply, are thus not only abstract financial figures, but concrete goods in the sense of products and services as artefacts of human activities in manufacturing and design. Manufacturing technology significantly determines how exactly humans create these artefacts, and thus how they shape their environment, communities and individual lives. Directing these human activities to coping with challenges of sustainability is, consequently, a relevant research contribution in manufacturing technology.

In the politically charged arena of sustainable manufacturing with its high economic impact and huge variety of conflicting interest groups, the comparatively neutral position of science can serve to help win over people's trust. At the same time, innovative approaches, methods and tools need to be scientifically developed in order to overcome the educational gap regarding sustainable development and even more sustainable manufacturing. The triangle of researching, educating and networking that determines schools' and universities' daily agendas likewise involves the three pillars of ESD science: researching and developing innovative didactic approaches (1), putting them into direct use by integrating them into education as awareness-raising activities (2), and making use of universities' unique localization as experts standing in between politics, industry and a great number of learners in pursuit of building networks for promoting ESD (3).

References

Ahnert, Liselotte. 2014. *Theorien in der Entwicklungspsychologie*. Heidelberg: Springer.
Bolscho, Dietmar. 2005. Der Beitrag der Erziehungswissenschaften für die Nachhaltigkeitskommunikation. In *Handbuch Nachhaltigkeits-kommunikation*, ed. Gerd Michelsen, Jasmin Godemann, 143–150. Munich: Oekom.
Bormann, Inka. 2005. Zwischen Wunsch und Wirklichkeit: Nachhaltigkeitskommunikation in Schulen. In *Handbuch Nachhaltigkeits-kommunikation*, ed. Gerd Michelsen, Jasmin Godemann, 793–803. Munich: Oekom.
Brand, Karl-Werner. 2005. Nachhaltigkeitskommunikation: eine soziologische Perspektive. In *Handbuch Nachhaltigkeitskommunikation*, ed. Gerd Michelsen, Jasmin Godemann, 151–161. Munich: Oekom.
Bormann, Inka, Kristina Broens, Felix Brümmer, Bettina Klaczinski, Markus Lindner, Carina Lücke, Sebastian Niedlich, and Sonja Warning. 2015. *Evaluation der Kampagne "Schule der Zukunft – Bildung für Nachhaltigkeit"*. Düsseldorf: Ministerium für Schule und Weiterbildung des Landes Nordrhein-Westfalen, and Ministerium für Klimaschutz, Umwelt, Landwirtschaft, Natur- und Verbraucherschutz des Landes Nordrhein-Westfalen. http://www.schule-der-zukunft.nrw.de/fileadmin/user_upload/Schule-der-Zukunft/Aktuelles/Abschlussbericht_Evaluation_Schule_der_Zukunft_ENDVERSION.pdf. Accessed 27 Sept 2016.
Chryssolouris, George. 2005. *Manufacturing systems: Theory and practice*. New York: Springer.
Crawley, Edward F., Johan Malmqvist, Sören Ostlund, and Doris R. Brodeur. 2007. *Rethinking engineering education: The CDIO approach*. New York: Springer.
Dave, Ravindrakumar H. 1970. Psychomotor levels. In *Developing and writing behavioral objectives*, ed. Robert J. Armstrong, 33–34. Tucson: Educational Innovators Press.
Fischer, Daniel, and Matthias Barth. 2014. Key competencies for and beyond sustainable consumption. An educational contribution to the debate. *GAIA* 23(S1): 193–200.
Fiske, Susan T., and Cydney Dupree. 2014. Gaining trust as well as respect in communicating to motivated audiences about science topics. *PNAS* 111(4): 13593–13597.
Gosling, David, and Jennifer A. Moon. 2001. *How to use learning outcomes and assessment criteria*. London: SEEC.
Griese, Kai-Michael, Christel Kumbruck, and Christin Michaelis. 2014. Der Einfluss von Humor auf die Nachhaltigkeitskommunikation. *Ökologisches Wirtschaften*, 29(4): 35–38.
Grunenberg, Heiko, and Udo Kuckartz. 2005. Umweltbewusstsein. Empirische Erkenntnisse und Konsequenzen für die Nachhaltigkeitskommunikation. In *Handbuch Nachhaltigkeitskommunikation*, ed. Gerd Michelsen, Jasmin Godemann, 195–206. Munich: Oekom.
Harrow, Anita J. 1972. *A taxonomy of the psychomotor domain: A guide for developing behavioral objectives*. New York: Longman.
Kirkpatrick, Donald. 1996. Great ideas revisited. Techniques for evaluating training programmes: Revisiting Kirkpatrick's four-level model. *Training and Development* 50(1): 54–59.

Kolb, David A. 1984. *Experiential learning: Experience as the source of learning and development*. Englewood Cliffs: Prentice Hall.

Krüger, Jörg, and The Duy Nguyen. 2015. Automated vision-based live ergonomics analysis in assembly operations. *CIRP Annals—Manufacturing Technology* 64(1): 9–12.

Kruse, Lenelis. 2005. Nachhaltigkeitskommunikation und mehr: die Perspektive der Psychologie. In *Handbuch Nachhaltigkeitskommunikation*, ed. Gerd Michelsen, and Jasmin Godemann, 109–120. Munich: Oekom.

Lasswell, Harold D. 1948. The structure and function of communication in society. In *The communication of ideas: A series of addresses*, ed. Lyman Bryson, 32–51. New York: Harper.

Leacock, Tracey L., and John C. Nesbit. 2007. A framework for evaluating the quality of multimedia learning resources. *Educational Technology & Society* 10(2): 44–59.

Manufuture High Level Group and Support Group. 2006. *Strategic research agenda: Assuring the future of manufacturing in Europe*. Brussels: ManuFuture Platform. https://www.um.es/operum/plataformas/files/manufuture.pdf. Accessed 27 Sept 2016.

Mattes, Wolfgang. 2011. *Methoden für den Unterricht: Kompakte Übersichten für Lehrende und Lernende*. Paderborn: Schöningh.

Mayer, Horst Otto, Johannes Hertnagel, and Heidi Weber. 2009. *Lernzielüberprüfung im eLearning*. München: Oldenbourg.

McFarland, Randy, Carsten Reise, Alexandra Postawa, and Günther Seliger. 2013. Learnstruments in value creation and learning centered work place design. In *Proceedings of the 11th global conference on sustainable manufacturing—innovative solutions*, ed. Günther Seliger, 624–629. Berlin: Universitätsverlag der TU Berlin.

Michelsen, Gerd. 2005. Nachhaltigkeitskommunikation: Verständnis – Entwicklung – Perspektiven. In *Handbuch Nachhaltigkeitskommunikation*, edited by Gerd Michelsen, Jasmin Godemann, 25–41. Munich: Oekom.

Muschard, Bernd, and Günther Seliger. 2015. Realization of a learning environment to promote sustainable value creation in areas with insufficient infrastructure. *Procedia CIRP* 32: 70–75.

Neugebauer, Sabrina, Ya-Ju Chang, Markus Maliszewski, Kai Lindow, Rainer Stark, and Matthias Finkbeiner. 2013. Life cycle sustainability assessment & sustainable product development: A case study on pedal electric cycles (Pedelec). In *Proceedings of the 11th global conference on sustainable manufacturing—innovative solutions*, ed. Günther Seliger, 549–554. Berlin: Universitätsverlag der TU Berlin.

Pappas, Eric, Olga Pierrakos, and Robert L. Nagel. 2013. Using Bloom's taxonomy to teach sustainability in multiple contexts. *Journal of Cleaner Production* 48: 54–64.

Reeves, Phil. 2012. Democratizing Manufacturing: 3D Printing in the Developing World. Yumpu. https://www.yumpu.com/en/document/view/46578172/democratizing-manufacturing-econolyst/1. Accessed 27 Sept 2016.

Roeder, Ina, Matthias Scheibleger, and Rainer Stark. 2015. How to make people make a change: Using social labelling for raising awareness on sustainable manufacturing. *Procedia CIRP* 40: 359–364.

Roeder, Ina, Mustafa Severengiz, Rainer Stark, and Günther Seliger. 2016. Open educational resources as driver for manufacturing-related education for learning of sustainable development. Paper to be presented at 14th Global Conference on Sustainable Manufacturing, Stellenbosch, South Africa, October 3–5, 2016.

Simpson, Elizabeth J. 1972. *The classification of educational objectives in the psychomotor domain*. Urbana: Gryphon House.

Seliger, Günther. 2012. Sustainable manufacturing for global value creation. In *Sustainable manufacturing: Shaping global value creation*, ed. Günther Seliger, 3–8. Berlin: Springer.

Seliger, Günther, Carsten Reise, and Rand McFarland. 2009. Outcome-oriented learning environment for sustainable engineering education. In *Proceedings of the 7th global conference on sustainable manufacturing—sustainable product development and life cycle engineering*, ed. Günther Seliger, 91–98. Berlin: Universitätsverlag der TU Berlin.

Tan, D., S. Amershi, B. Begole, W.A. Kellogg, M. Tungare, S. Deterding, et al. 2011. Gamification using game-design elements in non-gaming contexts. In *Proceedings of the 2011 annual conference extended abstracts on Human factors in computing systems*, ed. D. Tan, 2425–2428. New York, NY: ACM.

Ueda, Kanji, Takeshi Takenaka, Jozsef Váncza, and László Monostori. 2009. Value creation and decision-making in sustainable society. *CIRP Annals—Manufacturing Technology* 58(2): 681–700.

UNESCO. 2012. 2012 World Open Educational Resources Congress: Wednesday 20–Friday 22 June, 2012. http://www.unesco.org/new/fileadmin/MULTIMEDIA/HQ/CI/CI/pdf/CI_Information_Meetings/2012_world_oer_congress_en.pdf. Accessed 27 Sept 2016.

Wals, Arjen E.J. 2015. Social learning-oriented capacity-building for critical transitions towards sustainability. In *Schooling for sustainable development in Europe: Concepts, policies and educational experiences at the end of the UN decade of education for sustainable development*, ed. Rolf Jucker, and Reiner Mathar, 87–107. Cham: Springer.

Wang, Wei M., Lars Wolter, Kai Lindow, and Rainer Stark. 2014. Graphical visualization of sustainable manufacturing aspects for knowledge transfer to public audience. *Procedia CIRP* 26: 58–63.

Weintraub Austin, Erica, and Bruce E. Pinkleton. 2015. *Strategic public relations management: Planning and managing effective communication programs*. New York: Routledge.

Zirfas, Jörg, and Michael Göhlich. 2007. *Lernen: Ein pädagogischer Grundbegriff*. Stuttgart: Kohlhammer.

Glossary

Manufacturing

Jérémy Bonvoisin, Technische Universität Berlin, Institute for Machine tools and Factory Management, Chair of Industrial Information Technology.

No widely accepted and unambiguous definition of this term currently exists. Various definitions from encyclopaedia and international standards of reference feature rather blurry boundaries of the activities which manufacturing encompasses, along with an unclear overlap with the term *industry*. This goes hand in hand with an overall difficulty in finding consistent macroeconomic data on the manufacturing sector (such as energy consumption and greenhouse gas emissions). There are generally two ways of defining manufacturing: as an economic sector, i.e. by reference of the type of output it generates and input it requires, or as an organisation of value creation, i.e. by identifying the specific characteristics of manufacturing activities. The next subsections present these two views of the term in detail and suggest a definition for the purpose of consistency within this book.

Manufacturing as an Industrial Sector

The Encyclopaedia Britannica considers the term manufacturing in a broad sense as a synonym for "secondary industry"[1], i.e. the range of activities leading from raw materials to finished products, without distinction between continuous

---

[1] From the Encyclopaedia Britannica (accessed 16.02.2016): "Alternative title: secondary industry. Manufacturing, any industry that makes products from raw materials by the use of manual labour or machinery and that is usually carried out systematically with a division of labour. (See industry.) In a more limited sense, manufacturing denotes the fabrication or assembly of components into finished products on a fairly large scale. Among the most important manufacturing industries are those that produce aircraft, automobiles, chemicals, clothing, computers, consumer electronics, electrical equipment, furniture, heavy machinery, refined petroleum products, ships, steel, and tools and dies. [...]"

© The Editor(s) (if applicable) and The Author(s) 2017
R. Stark et al. (eds.), *Sustainable Manufacturing*, Sustainable Production, Life Cycle Engineering and Management, DOI 10.1007/978-3-319-48514-0

(e.g. "chemicals" and "refined petroleum products") and discrete products (e.g. "automobiles" and "heavy machinery"). The free encyclopaedia Wikipedia delivers a similar definition[2] focusing on the production of finished products, wherein components being integrated into other products are also considered to be finished products and thus likewise as input materials of manufacturing activities. These definitions however conflate the horizontal (range of products) and vertical (to what degree products are "finished") boundaries of the activities embraced. Recognizing the difficulty in delivering a positive definition of the sector, the International Standard Industrial Classification of All Economic Activities (revision 4) of the United Nations Statistics Division delivers an ad hoc and negative definition of manufacturing, i.e. it avoids defining what is the common denominator of these activities, but instead gives a list of what products are considered to be the product of manufacturing and what products do *not* fall into this category. This definition leaves out naming the full scope of manufacturing activities such as mining or energy and water supply, while however including mention of the production of a large and heterogeneous range of continuous and discrete finished products such as food beverages, textiles, petroleum products or furniture. A similar definition is also available for the European area (Statistical classification of economic activities in the European Community, abbreviated as NACE).

Manufacturing as an Organisation of Value Creation

According to the Encyclopaedia Britannica, manufacturing is characterised by the "fabrication or assembly of components into finished products on a fairly large scale." The free encyclopaedia Wikipedia also considers the scale of production as a determining factor and adds that manufacturing is performed through the "use or sale using labour and machine, tools, chemical and biological processing." It however puts the concept of large scale into perspective by indicating that manufacturing activities may range from handicrafts to high tech enterprise. The Columbia Encyclopaedia indicates on this point that manufacturing is not to be juxtaposed against handcraft activities, since manufacturing indeed forms the basis of craftsmanship. However, apart from the question of scale, the concentration of production factors and the departure from handiwork, the defining characteristics of

---

[2]From the free encyclopedia Wikipedia (accessed 16.02.2016): "Manufacturing is the production of merchandise for use or sale using labour and machines, tools, chemical and biological processing, or formulation. The term may refer to a range of human activity, from handicraft to high tech, but is most commonly applied to industrial production, in which raw materials are transformed into finished goods on a large scale. Such finished goods may be used for manufacturing other, more complex products, such as aircraft, household appliances or automobiles, or sold to wholesalers, who in turn sell them to retailers, who then sell them to end users – the `consumers'. [...] Modern manufacturing includes all intermediate processes required for the production and integration of a product's components. Some industries, such as semiconductor and steel manufacturers use the term fabrication instead. [...]"

manufacturing in this encyclopaedia are identified as the capitalistic organisation and the division of labour.

Definition in the Context of This Book

In pursuit of a narrowing down of the scope of the term to activities that are in the focus of the competencies of the authors, in the context of this book, we have decided to define manufacturing as the range of activities contributing to the fabrication or assembly of raw materials and components into discrete finished products through systematic division of labour. This term covers all necessary activities allowing the execution of manufacturing operations, i.e. design and operation of physical processes (e.g. machining), corresponding overhead processes (such as compressed air generation) and surrounding organisational processes such as product development, factory planning and supply chain management.

This definition covers both manner of considering manufacturing: as a subset of the industrial sector and as an organisation of value creation. It excludes the production of continuous products from the process industry by the reference to *discrete products*, defined by Duflou et al. (2012) as outputs that "can be identified and [are] measurable in distinct units rather than by weight or volume as in process industry." These authors underline that process industries have been under focus for many years, as they represent the largest share of energy consumption among all industries. Aluminium production, for instance, released in 2009 around 1% of the global annual anthropic greenhouses gases (Liu et al. 2013). The definition adopted also excludes activities from craftsmanship by mentioning the systematic division of labour.

Sustainability

Ya-Ju Chang (ya-ju.chang@tu-berlin.de), Sabrina Neugebauer, Annekatrin Lehmann, René Scheumann and Matthias Finkbeiner, Chair of Sustainable Engineering, Department of Environmental Technology, Technische Universität Berlin.

The term "sustainability" is a concept that emphasizes the balanced continuity of nature and human society (Vehkamäki 2005). It has raised public awareness and become a high-profile topic increasingly preoccupying government entities and industries.

Sustainability as a term has been used since the late Middle Ages and was originally coined in connection with sustainable forest management, where it surfaced in the Saxon Forest Regulation in 1560 (Augusti 1839). It was further mentioned by Carl von Carlowitz in the book published in 1713 in the context of his proposal that continuous, permanent and sustainable utilisation become the rule for forestry (von Carlowitz 1713). The report "Limits of growth" of the Club of Rome in 1972 (Meadows et al. 1972) reintroduced the word in connection with human development and considers the limits of available resources, damages to the natural environment, and poverty of human societies. In 1987, the report known as "Our Common Future" by the Brundtland commission referred to sustainable development, as a development that "meets the needs of the present generation without compromising the ability of future generations to meet their own needs" (Brundtland et al. 1987). In other words, "sustainable development would create and maintain the conditions under which humans and nature can exist in productive harmony that permits fulfilling the social, economic and other requirements of present and future generations"[3].

Following these definitions, sustainability is considered within this book consisting of three dimensions: environment, society, and economy, as displayed by the following figure.

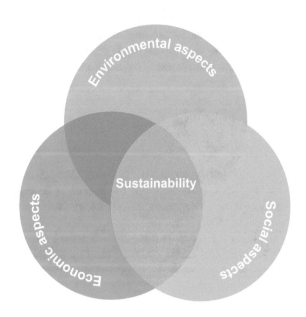

---

[3]US-EPA: http://www.epa.gov/sustainability/basicinfo.htm#sustainability (access June 2013).

Sustainable Manufacturing

Jérémy Bonvoisin, Technische Universität Berlin, Institute for Machine tools and Factory Management, Chair of Industrial Information Technology.

Authors from the field seem only to agree on the fact that no widely accepted definition of the term sustainable manufacturing currently exists (Jayal et al. 2010; see for example Haapala et al. 2013). It therefore unfortunately inherits doubly confounded fuzziness from the definitions of both manufacturing and sustainability combined—both terms being defined by this glossary. The next subsections present a review of existing definitions of the literature and suggest a definition for the sake of consistency within this book.

Scope of Existing Definitions

Existing definitions collected from the literature show mostly divergences regarding the explicit inclusion of the social dimension of sustainability. While some papers provide definitions referring only to the environmental dimension of sustainability (implicitly including the economic dimension), some others refer to the three dimensions. This observation corroborates the statement of Haapala et al. (2013) who notes that "sustainable manufacturing is sometimes used carelessly to describe the actions related to characterizing and reducing the environmental impacts of manufacturing." However, as noted by Mihelcic et al. (2003), solutions focusing only on the environmental solutions are insufficient since "even systems with efficient material and energy use can overwhelm the carrying capacity of a region or lead to other socially unacceptable outcomes."

Among the definitions addressing all three dimensions of sustainability, the one used the most is that of the United States Department of Commerce (DOC) and the Environmental Protection Agency (EPA). It defines sustainable manufacturing as "the creation of manufactured products through economically-sound processes that minimize negative environmental impacts while conserving energy and natural resources. Sustainable manufacturing also protects employee, community, and consumer safety." Another often cited definition from Mihelcic et al. (2003) defines sustainable manufacturing as the "design of human and industrial systems to ensure that humankind's use of natural resources and cycles do not lead to diminished quality of life due either to losses in future economic opportunities or to adverse impacts on social conditions, human health, and the environment." Both definitions implicitly consider all three dimensions of sustainability.

Definition in the Context of This Book

Following the choice of considering all dimensions of sustainability, we have adopted a definition close to that of the EPA/DOC, but which reinforces the integration of the three dimensions while however leaving the concretization of sustainability dimensions open. Sustainable manufacturing is defined here as *the creation of discrete manufactured products that, in fulfilling their functionality over their entire life cycle, cause a manageable amount of impacts on the environment (nature and society) whilst delivering economic and societal value.*

Note that environmental engineering, i.e. a range of engineering and managerial techniques concerned with the protection of the environmental quality of a given area, is <u>not</u> included in the scope of this definition. Approaches such as solid waste management, water supply, wastewater treatment, air pollution management or even geoengineering are therefore not addressed in this contribution.

References

Augusti, C. 1839. Forst- und Holz-Ordnung zu Sachsen, den 8. Sept. 1560. In *Handbuch aller seit 1560 bis auf die neueste Zeit erschienenen Forst- und Jagdgesetze des Königreichs Sachsen : systematisch und chronologisch zusammengestellt*, ed. by G.V. Schmid, 405. 1, Forst-Gesetze. Meißen: Goedsche.

Brundtland, G., M. Khalid, S. Agnelli, et al. 1987. *Our common future*. Oxford University Press.

Duflou, J.R., J.W. Sutherland, D. Dornfeld, et al. 2012. Towards energy and resource efficient manufacturing: A processes and systems approach. *CIRP Annals—Manufacturing Technology* 61: 587–609. doi:10.1016/j.cirp.2012.05.002.

Haapala, K.R., F. Zhao, J. Camelio, et al. 2013. A review of engineering research in sustainable manufacturing. *Journal of Manufacturing Science and Engineering* 135: 041013–041013. doi:10.1115/1.4024040.

Jayal, A.D., F. Badurdeen, O.W. Dillon Jr., and I.S. Jawahir. 2010. Sustainable manufacturing: Modeling and optimization challenges at the product, process and system levels. *CIRP Journal of Manufacturing Science and Technology* 2: 144–152. doi:10.1016/j.cirpj.2010.03.006.

Liu, G., C.E. Bangs, and D.B. Müller. 2013. Stock dynamics and emission pathways of the global aluminium cycle. *Nature Climate Change* 3: 338–342. doi:10.1038/nclimate1698.

Meadows, D.H., D.L. Meadows, J. Randers, and W.W. Behrens. 1972. *Limits to growth*. Universe Books.

Mihelcic, J.R., J.C. Crittenden, M.J. Small, et al. 2003. Sustainability science and engineering: The emergence of a new metadiscipline. *Environmental Science & Technology* 37: 5314–5324. doi:10.1021/es034605h.

Vehkamäki, S. 2005. The concept of sustainability in modern times. Sustainable use of renewable natural resources—from principles to practices. University of Helsinki, Department of Forest. *Ecology Publications* 34: 13.

Von Carlowitz, H.C. 1713. *Sylvicultura Oeconomica oder haußwirthliche Nachricht und Naturgemäßige Anweisung zur Wilden Baum-Zucht*. Leipzig: Johann Friedrich Braun Erben.

Printed in the United States
By Bookmasters